DELMAR's Standard Textbook of Electricity

Fifth Edition

Volume 3
Ontario Electrical Apprenticeship Program

Stephen L. Herman

NELSON EDUCATION

COPYRIGHT © 2014 by Nelson Education Ltd.

Printed and bound in Canada
1 2 3 4 16 15 14 13

For more information contact Nelson Education Ltd., 1120 Birchmount Road, Toronto, Ontario, M1K 5G4. Or you can visit our Internet site at http://www.nelson.com

ALL RIGHTS RESERVED. No part of this work covered by the copyright herein may be reproduced, transcribed, or used in any form or by any means—graphic, electronic, or mechanical, including photocopying, recording, taping, Web distribution, or information storage and retrieval systems—without the written permission of the publisher.

For permission to use material from this text or product, submit all requests online at www.cengage.com/permissions. Further questions about permissions can be emailed to permissionrequest@cengage.com

Every effort has been made to trace ownership of all copyrighted material and to secure permission from copyright holders. In the event of any question arising as to the use of any material, we will be pleased to make the necessary corrections in future printings.

This textbook is a Nelson custom publication. Because your instructor has chosen to produce a custom publication, you pay only for material that you will use in your course.

ISBN-13: 978-0-17-665680-5
ISBN-10: 0-17-665680-4

Consists of Selections from:

Delmar's Standard Textbook of Electricity, *5th Edition*, *Herman*
ISBN 1-111-53915-4, © 2011, 2009
Delmar, Cengage Learning

Cover Credit:
Joe Belanger/Shutterstock

Contents

SECTION XI
The Wattmeter — 722

UNIT 25
Measuring Instruments — 723

25–1 The Wattmeter — 724

SECTION XII
Three-Phase Power — 728

UNIT 26
Three-Phase Circuits — 729

26–1 Three-Phase Circuits — 730
26–2 Wye Connections — 732
26–3 Delta Connections — 736
26–4 Three-Phase Power — 737
26–5 Watts and VARs — 738
26–6 Three-Phase Circuit Calculations — 739
26–7 Load 3 Calculations — 747
26–8 Load 2 Calculations — 748
26–9 Load 1 Calculations — 749
26–10 Alternator Calculations — 749
26–11 Power Factor Correction — 750

SECTION XIII
Transformers — 758

UNIT 27
Single-Phase Transformers — 759

27–1 Single-Phase Transformers — 760
27–2 Isolation Transformers — 762
27–3 Autotransformers — 788
27–4 Transformer Polarities — 791
27–5 Voltage and Current Relationships in a Transformer — 796
27–6 Testing the Transformer — 798

27–7	Transformer Nameplates	799
27–8	Determining Maximum Current	800
27–9	Transformer Impedance	801

UNIT 28
Three-Phase Transformers — 816

28–1	Three-Phase Transformers	817
28–2	Closing a Delta	822
28–3	Three-Phase Transformer Calculations	823
28–4	Open-Delta Connection	829
28–5	Single-Phase Loads	830
28–6	Closed Delta with Center Tap	834
28–7	Closed Delta without Center Tap	835
28–8	Delta–Wye Connection with Neutral	836
28–9	T-Connected Transformers	837
28–10	Scott Connection	840
28–11	Zig-Zag Connection	840
28–12	Harmonics	842

SECTION XIV
AC Machines — 856

UNIT 29
Three-Phase Alternators — 857

29–1	Three-Phase Alternators	858
29–2	The Rotor	862
29–3	The Brushless Exciter	862
29–4	Alternator Cooling	865
29–5	Frequency	866
29–6	Output Voltage	867
29–7	Paralleling Alternators	868
29–8	Sharing the Load	871
29–9	Field-Discharge Protection	871

UNIT 30
Three-Phase Motors — 875

30–1	Three-Phase Motors	876
30–2	The Rotating Magnetic Field	877
30–3	Connecting Dual-Voltage Three-Phase Motors	889
30–4	Squirrel-Cage Induction Motors	895
30–5	Wound-Rotor Induction Motors	914
30–6	Synchronous Motors	917
30–7	Selsyn Motors	923

UNIT 31
Single-Phase Motors — 931

31–1	Single-Phase Motors	932
31–2	Split-Phase Motors	932
31–3	Resistance-Start Induction-Run Motors	936
31–4	Capacitor-Start Induction-Run Motors	944
31–5	Dual-Voltage Split-Phase Motors	946
31–6	Determining the Direction of Rotation for Split-Phase Motors	949
31–7	Capacitor-Start Capacitor-Run Motors	950
31–8	Shaded-Pole Induction Motors	953
31–9	Multispeed Motors	957
31–10	Repulsion-Type Motors	959
31–11	Construction of Repulsion Motors	959
31–12	Repulsion-Start Induction-Run Motors	963
31–13	Repulsion-Induction Motors	965
31–14	Single-Phase Synchronous Motors	966
31–15	Stepping Motors	969
31–16	Universal Motors	977

APPENDIX F
Answers to Practice Problems — 986

Index — 989

XI The Wattmeter

Unit 25
Measuring Instruments

Why You Need to Know

Measuring instruments are the eyes of the electrician. An understanding of how measuring instruments operate is very important to anyone working in the electrical field. They provide the electrician with the ability to evaluate problems on the job through the use of technical tools. They also enable an electrician to correctly determine electrical values of voltage, current, resistance, power, and many others. In this unit you will learn

- the differences between dynamic and electronic wattmeters.

OUTLINE

25–1 The Wattmeter

KEY TERMS

Wattmeter

SECTION XI The Wattmeter

Objectives

After studying this unit, you should be able to

- discuss the differences between dynamic and electronic wattmeters.
- connect a wattmeter into a circuit.

Preview

Anyone desiring to work in the electrical and electronics field must become proficient with the common instruments used to measure electrical quantities. The wattmeter is an instrument for measuring electric power or watts in a given circuit. ■

25–1 The Wattmeter

The **wattmeter** is used to measure true power in a circuit. There are two basic types of wattmeters, dynamic and electronic. Dynamic wattmeters differ from d'Arsonval-type meters in that they do not contain a permanent magnet. They contain an electromagnet and a moving coil *(Figure 25–1)*. The electromagnets

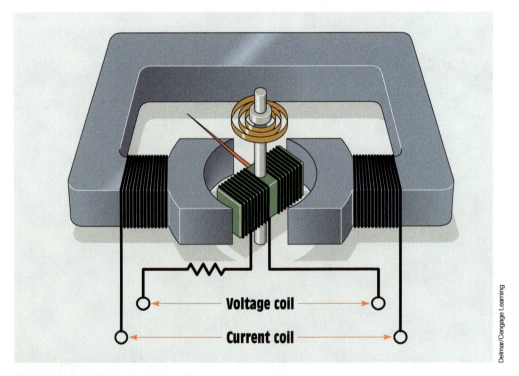

FIGURE 25-1 The wattmeter contains two coils—one for voltage and the other for current.

are connected in series with the load in the same manner that an ammeter is connected. The moving coil has resistance connected in series with it and is connected directly across the power source in the same manner as a voltmeter *(Figure 25–2)*.

Because the electromagnet is connected in series with the load, the current flow through the load determines the magnetic field strength of the stationary magnet. The magnetic field strength of the moving coil is determined by the amount of line voltage. The turning force of the coil is proportional to the strength of these two magnetic fields. The deflection of the meter against the spring is proportional to the amount of current flow and voltage.

Because the wattmeter contains an electromagnet instead of a permanent magnet, the polarity of the magnetic field is determined by the direction of current flow. The same is true of the polarity of the moving coil connected across the source of voltage. If the wattmeter is connected into an AC circuit, the polarity of the two coils will reverse at the same time, producing a continuous torque. For this reason, the wattmeter can be used to measure power in either a DC or an AC circuit. However, if the connection of the stationary coil or the moving coil is reversed, the meter will attempt to read backward.

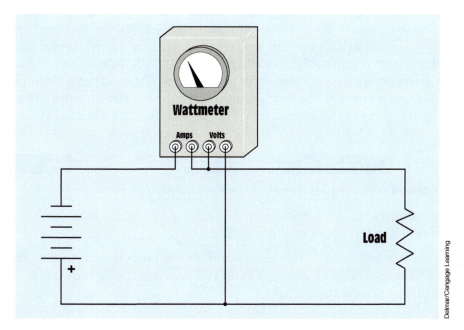

FIGURE 25-2 The current section of the wattmeter is connected in series with the load, and the voltage section is connected in parallel with the load.

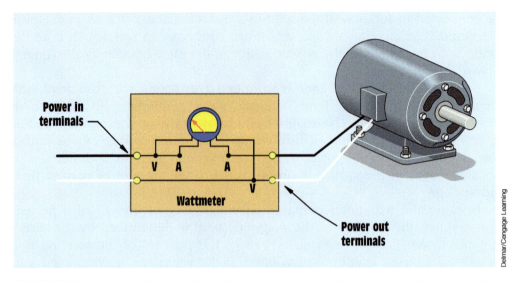

FIGURE 25–3 Portable wattmeters often make connection to the voltage and current terminals inside the meter.

Dynamic-type wattmeters are being replaced by wattmeters that contain electronic circuitry to determine true power. They are less expensive and generally more accurate than the dynamic type. Like dynamic wattmeters, electronic-type meters contain amperage terminals that connect in series with the load and voltage terminals that connect in parallel with the load. Portable-type wattmeters often have terminals labeled "power in" and "power out." Connection to the current and voltage section of the meter is made inside the meter *(Figure 25-3)*. Analog-type electronic wattmeters use a standard d'Arsonval-type movement to indicate watts. The electronic circuit determines the true power of the circuit and then supplies the appropriate power to the meter movement. Wattmeters with digital displays are also available.

Summary

- Wattmeters contain a stationary coil and a movable coil.
- The stationary coil of a wattmeter is connected in series with the load, and the moving coil is connected to the line voltage.
- The turning force of the dynamic wattmeter is proportional to the strength of the magnetic field of the stationary coil and the strength of the magnetic field of the moving coil.

XII Three-Phase Power

Unit 26
Three-Phase Circuits

OUTLINE

26–1 Three-Phase Circuits
26–2 Wye Connections
26–3 Delta Connections
26–4 Three-Phase Power
26–5 Watts and VARs
26–6 Three-Phase Circuit Calculations
26–7 Load 3 Calculations
26–8 Load 2 Calculations
26–9 Load 1 Calculations
26–10 Alternator Calculations
26–11 Power Factor Correction

KEY TERMS

Delta connection
Line current
Line voltage
Phase current
Phase voltage
Star connection
Three-phase VARs
Three-phase watts
Wye connection

Why You Need to Know

Most of the power produced in the world is three phase. The importance of understanding the principles concerning three-phase power cannot be overstated. Three-phase power is used to run almost all industry throughout the United States and Canada. Even residential power is derived from a three-phase power system. This unit

- discusses the basics of three-phase power generation as the most common electric power source in the world.
- explains how three-phase power is produced and the basic connections used by devices intended to operate on three-phase power.
- illustrates how to calculate voltage, current, and power for different types of three-phase loads.
- explains why electricians must be able to understand and calculate the different voltages in a three-phase circuit.
- determines the power factor and how to correct it in a three-phase system.

SECTION XII Three-Phase Power

Objectives

After studying this unit, you should be able to

- discuss the differences between three-phase and single-phase voltages.
- discuss the characteristics of delta and wye connections.
- calculate voltage and current values for delta and wye circuits.
- connect delta and wye circuits and make measurements with measuring instruments.
- calculate the amount of capacitance needed to correct the power factor of a three-phase motor.

Preview

Most of the electric power generated in the world today is three phase. Three-phase power was first conceived by Nikola Tesla. In the early days of electric power generation, Tesla not only led the battle concerning whether the nation should be powered with low-voltage DC or high-voltage AC, but he also proved that three-phase power was the most efficient way that electricity could be produced, transmitted, and consumed. ■

26–1 Three-Phase Circuits

There are several reasons why three-phase power is superior to single-phase power:

1. The horsepower rating of three-phase motors and the kilovolt-ampere rating of three-phase transformers are about 150% greater than for single-phase motors or transformers with a similar frame size.

2. The power delivered by a single-phase system pulsates *(Figure 26–1)*. The power falls to zero three times during each cycle. The power delivered by a three-phase circuit pulsates also, but it never falls to zero *(Figure 26–2)*. In a three-phase system, the power delivered to the load is the same at any instant. This produces superior operating characteristics for three-phase motors.

3. In a balanced three-phase system, the conductors need be only about 75% the size of conductors for a single-phase, two-wire system of the same kilovolt-ampere (kVA) rating. This savings helps offset the cost of supplying the third conductor required by three-phase systems.

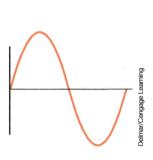

FIGURE 26–1 Single-phase power falls to zero two times each cycle.

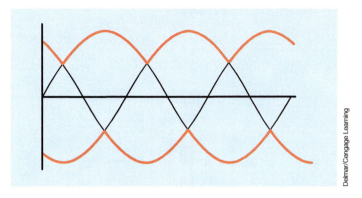

FIGURE 26–2 Three-phase power never falls to zero.

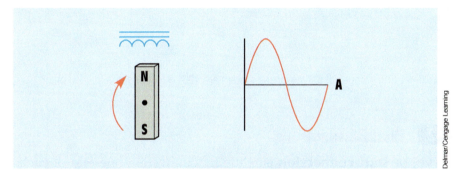

FIGURE 26–3 Producing a single-phase voltage.

A single-phase alternating voltage can be produced by rotating a magnetic field through the conductors of a stationary coil, as shown in *Figure 26–3*.

Because alternate polarities of the magnetic field cut through the conductors of the stationary coil, the induced voltage changes polarity at the same speed as the rotation of the magnetic field. The alternator shown in *Figure 26–3* is single phase because it produces only one AC voltage.

If three separate coils are spaced 120° apart, as shown in *Figure 26–4*, three voltages 120° out of phase with each other are produced when the magnetic field cuts through the coils. This is the manner in which a three-phase voltage is produced. There are two basic three-phase connections: the wye, or star, and the delta.

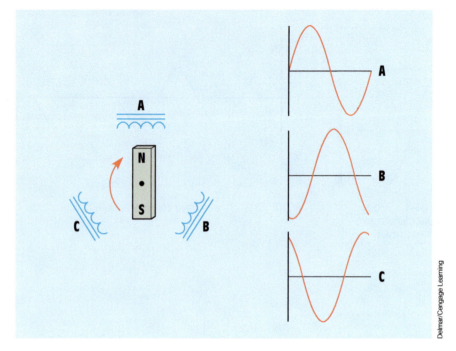

FIGURE 26–4 The voltages of a three-phase system are 120° out of phase with each other.

26–2 Wye Connections

The **wye,** or **star, connection** is made by connecting one end of each of the three-phase windings together *(Figure 26–5)*. The voltage measured across a single winding, or phase, is known as the **phase voltage** *(Figure 26–6)*. The voltage measured between the lines is known as the line-to-line voltage, or simply as the **line voltage.**

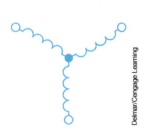

FIGURE 26–5 A wye connection is formed by joining one end of each of the windings together.

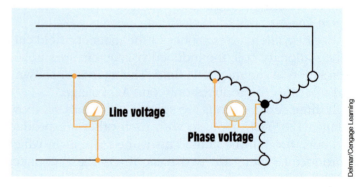

FIGURE 26–6 Line and phase voltages are different in a wye connection.

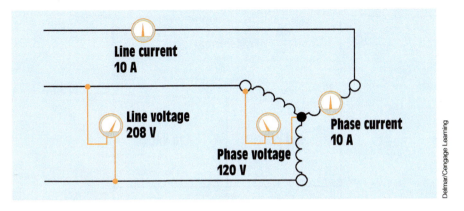

FIGURE 26-7 Line current and phase current are the same in a wye connection.

In *Figure 26-7*, ammeters have been placed in the phase winding of a wye-connected load and in the line that supplies power to the load. Voltmeters have been connected across the input to the load and across the phase. A line voltage of 208 volts has been applied to the load. Notice that the voltmeter connected across the lines indicates a value of 208 volts, but the voltmeter connected across the phase indicates a value of 120 volts.

In a wye-connected system, the line voltage is higher than the phase voltage by a factor of the square root of 3 (1.732). Two formulas used to calculate the voltage in a wye-connected system are

$$E_{Line} = E_{Phase} \times 1.732$$

and

$$E_{Phase} = \frac{E_{Line}}{1.732}$$

Notice in *Figure 26-7* that 10 amperes of current flow in both the phase and the line. ***In a wye-connected system,*** **phase current** ***and*** **line current** ***are the same:***

$$I_{Line} = I_{Phase}$$

Voltage Relationships in a Wye Connection

Many students of electricity have difficulty at first understanding why the line voltage of the wye connection used in this illustration is 208 volts instead of 240 volts. Because line voltage is measured across two phases that have a voltage of 120 volts each, it would appear that the sum of the two voltages should be 240 volts. One cause of this misconception is that many students are familiar with the 240/120-volt connection supplied to most homes. If voltage is measured across

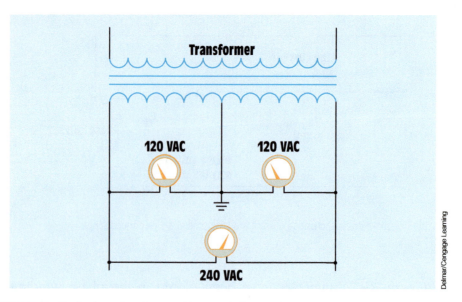

FIGURE 26–8 Single-phase transformer with grounded center tap.

the two incoming lines, a voltage of 240 volts will be seen. If voltage is measured from either of the two lines to the neutral, a voltage of 120 volts will be seen. The reason for this is that this is a single-phase connection derived from the center tap of a transformer *(Figure 26–8)*. The center tap is the midpoint of two out-of-phase voltages *(Figure 26–9)*. The vector sum of these two voltages is 240 volts.

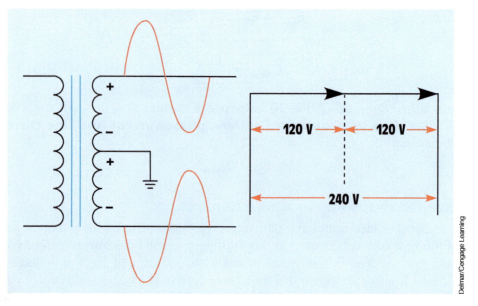

FIGURE 26–9 The voltages of a single-phase residential system are out of phase with each other.

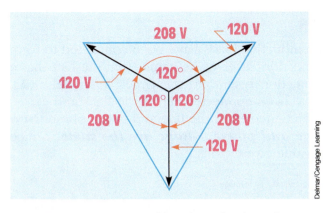

FIGURE 26–10 Vector sum of the voltages in a three-phase wye connection.

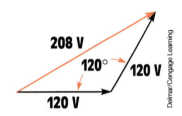

FIGURE 26–11 Adding voltage vectors of two-phase voltage values.

Three-phase voltages are 120° out of phase with each other, not in phase with each other. If the three voltages are drawn 120° apart, it will be seen that the vector sum of these voltages is 208 volts *(Figure 26–10)*. Another illustration of vector addition is shown in *Figure 26–11*. In this illustration, two-phase voltage vectors are added, and the resultant is drawn from the starting point of one vector to the end point of the other. The parallelogram method of vector addition for the voltages in a wye-connected three-phase system is shown in *Figure 26–12*.

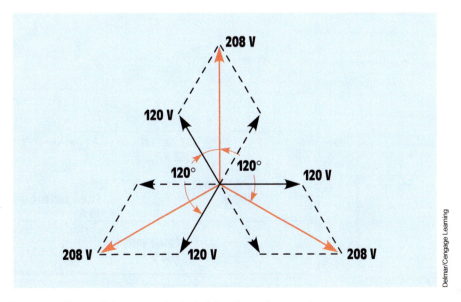

FIGURE 26–12 The parallelogram method of adding three-phase vectors.

26–3 Delta Connections

In *Figure 26–13*, three separate inductive loads have been connected to form a **delta connection.** This connection receives its name from the fact that a schematic diagram of this connection resembles the Greek letter delta (Δ). In *Figure 26–14*, voltmeters have been connected across the lines and across the phase. Ammeters have been connected in the line and in the phase. *In a delta connection, line voltage and phase voltage are the same.* Notice that both voltmeters indicate a value of 480 volts.

$$E_{Line} = E_{Phase}$$

The line current and phase current, however, are different. *The line current of a delta connection is higher than the phase current by a factor of the square root of 3 (1.732).* In the example shown, it is assumed that each of the phase windings has a current flow of 10 amperes. The current in each of the lines, however, is 17.32 amperes. The reason for this difference in current is that current flows through different windings at different times in a three-phase circuit. During some periods of time, current will flow between two lines only. At other times, current will flow from two lines to the third *(Figure 26–15)*. The delta connection is similar to a parallel connection because there is always more than one path for current flow. Because these currents are 120° out of phase with each other, vector addition must be used when finding the sum of the currents *(Figure 26–16)*. Formulas for determining the current in a delta connection are

$$I_{Line} = I_{Phase} \times 1.732$$

and

$$I_{Phase} = \frac{I_{Line}}{1.732}$$

FIGURE 26–13 Three-phase delta connection.

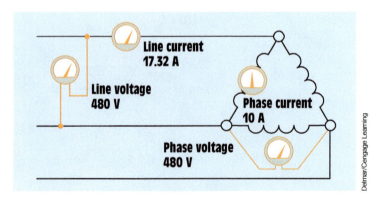

FIGURE 26–14 Voltage and current relationships in a delta connection.

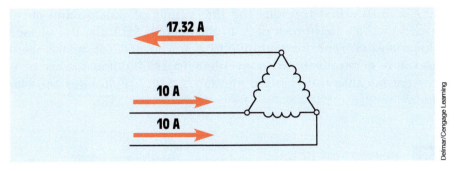

FIGURE 26–15 Division of currents in a delta connection.

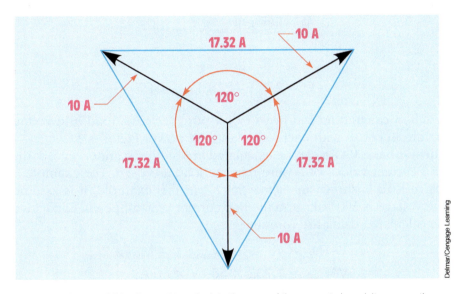

FIGURE 26–16 Vector addition is used to calculate the sum of the currents in a delta connection.

26-4 Three-Phase Power

Students sometimes become confused when calculating values of power in three-phase circuits. One reason for this confusion is that there are actually two formulas that can be used. If *line* values of voltage and current are known, the apparent power of the circuit can be calculated using the formula

$$VA = \sqrt{3} \times E_{Line} \times I_{Line}$$

If the *phase* values of voltage and current are known, the apparent power can be calculated using the formula

$$VA = 3 \times E_{Phase} \times I_{Phase}$$

Notice that in the first formula, the line values of voltage and current are multiplied by the square root of 3. In the second formula, the phase values of voltage and current are multiplied by 3. The first formula is used more because it is generally more convenient to obtain line values of voltage and current because they can be measured with a voltmeter and clamp-on ammeter.

26–5 Watts and VARs

Watts and VARs can be calculated in a similar manner. **Three-phase watts** can be calculated by multiplying the apparent power by the power factor:

$$P = \sqrt{3} \times E_{Line} \times I_{Line} \times PF$$

or

$$P = 3 \times E_{Phase} \times I_{Phase} \times PF$$

Note: When calculating the power of a pure resistive load, the voltage and current are in phase with each other and the power factor is 1.

Three-phase VARs can be calculated in a similar manner, except that voltage and current values of a pure reactive load are used. For example, a pure capacitive load is shown in *Figure 26–17*. In this example, it is assumed that the line voltage is 560 volts and the line current is 30 amperes. Capacitive VARs can be calculated using the formula

$$VARs_C = \sqrt{3} \times E_{Line\ (Capacitive)} \times I_{Line\ (Capacitive)}$$
$$VARs_C = 1.732 \times 560\ V \times 30\ A$$
$$VARs_C = 29{,}097.6$$

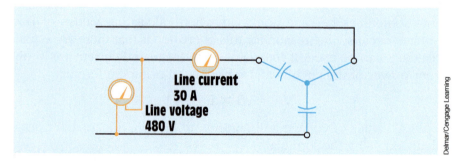

FIGURE 26–17 Pure capacitive three-phase load.

26–6 Three-Phase Circuit Calculations

In the following examples, values of line and phase voltage, line and phase current, and power are calculated for different types of three-phase connections.

■ EXAMPLE 26-1

A wye-connected three-phase alternator supplies power to a delta-connected resistive load *(Figure 26–18)*. The alternator has a line voltage of 480 volts. Each resistor of the delta load has 8 ohms of resistance. Find the following values:

$E_{L(Load)}$ — line voltage of the load

$E_{P(Load)}$ — phase voltage of the load

$I_{P(Load)}$ — phase current of the load

$I_{L(Load)}$ — line current to the load

$I_{L(Alt)}$ — line current delivered by the alternator

$I_{P(Alt)}$ — phase current of the alternator

$E_{P(Alt)}$ — phase voltage of the alternator

P — true power

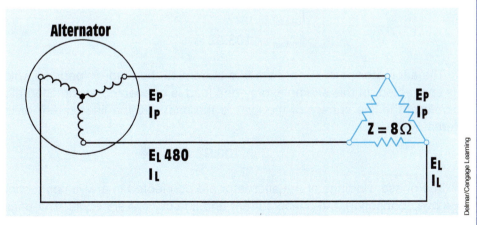

FIGURE 26–18 Calculating three-phase values using a wye-connected power source and a delta-connected load (example circuit).

Solution

The load is connected directly to the alternator. Therefore, the line voltage supplied by the alternator is the line voltage of the load:

$$E_{L(Load)} = 480 \text{ V}$$

The three resistors of the load are connected in a delta connection. In a delta connection, the phase voltage is the same as the line voltage:

$$E_{P(Load)} = E_{L(Load)}$$
$$E_{P(Load)} = 480 \text{ V}$$

Each of the three resistors in the load is one phase of the load. Now that the phase voltage is known (480 V), the amount of phase current can be calculated using Ohm's law:

$$I_{P(Load)} = \frac{E_{P(Load)}}{Z}$$

$$I_{P(Load)} = \frac{480 \text{ V}}{8 \text{ }\Omega}$$

$$I_{P(Load)} = 60 \text{ V}$$

The three load resistors are connected as a delta with 60 A of current flow in each phase. The line current supplying a delta connection must be 1.732 times greater than the phase current:

$$I_{L(Load)} = I_{P(Load)} \times 1.732$$
$$I_{L(Load)} = 60 \text{ A} \times 1.732$$
$$I_{L(Load)} = 103.92 \text{ A}$$

The alternator must supply the line current to the load or loads to which it is connected. In this example, only one load is connected to the alternator. Therefore, the line current of the load is the same as the line current of the alternator:

$$I_{L(Alt)} = 103.92 \text{ A}$$

The phase windings of the alternator are connected in a wye connection. In a wye connection, the phase current and line current are equal. The phase current of the alternator is therefore the same as the alternator line current:

$$I_{P(Alt)} = 103.92 \text{ A}$$

The phase voltage of a wye connection is less than the line voltage by a factor of the square root of 3. The phase voltage of the alternator is

$$E_{P(Alt)} = \frac{E_{L(Alt)}}{1.732}$$

$$E_{P(Alt)} = \frac{480 \text{ V}}{1.732}$$

$$E_{P(Alt)} = 277.136 \text{ V}$$

In this circuit, the load is pure resistive. The voltage and current are in phase with each other, which produces a unity power factor of 1. The true power in this circuit is calculated using the formula

$$P = 1.732 \times E_{L(Alt)} \times I_{L(Alt)} \times PF$$

$$P = 1.732 \times 480 \text{ V} \times 103.92 \text{ A} \times 1$$

$$P = 86,394.931 \text{ W}$$

EXAMPLE 26-2

A delta-connected alternator is connected to a wye-connected resistive load *(Figure 26–19)*. The alternator produces a line voltage of 240 V and the resistors have a value of 6 Ω each. Find the following values:

$E_{L(Load)}$ — line voltage of the load

$E_{P(Load)}$ — phase voltage of the load

$I_{P(Load)}$ — phase current of the load

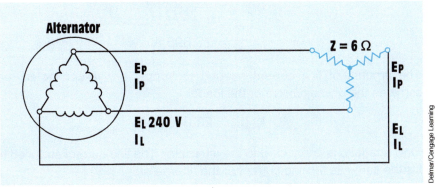

FIGURE 26–19 Calculating three-phase values using a delta-connected source and a wye-connected load (example circuit).

$I_{L(Load)}$ — line current to the load

$I_{L(Alt)}$ — line current delivered by the alternator

$I_{P(Alt)}$ — phase current of the alternator

$E_{P(Alt)}$ — phase voltage of the alternator

P — true power

Solution

As was the case in the previous example, the load is connected directly to the output of the alternator. The line voltage of the load must therefore be the same as the line voltage of the alternator:

$$E_{L(Load)} = 240 \text{ V}$$

The phase voltage of a wye connection is less than the line voltage by a factor of 1.732:

$$E_{P(Load)} = \frac{240 \text{ V}}{1.732}$$

$$E_{P(Load)} = 138.568 \text{ V}$$

Each of the three 6-Ω resistors is one phase of the wye-connected load. Because the phase voltage is 138.568 V, this voltage is applied to each of the three resistors. The amount of phase current can now be determined using Ohm's law:

$$I_{P(Load)} = \frac{E_{P(Load)}}{Z}$$

$$I_{P(Load)} = \frac{138.568 \text{ V}}{6 \text{ }\Omega}$$

$$I_{P(Load)} = 23.095 \text{ A}$$

The amount of line current needed to supply a wye-connected load is the same as the phase current of the load:

$$I_{L(Load)} = 23.095 \text{ A}$$

Only one load is connected to the alternator. The line current supplied to the load is the same as the line current of the alternator:

$$I_{L(Alt)} = 23.095 \text{ A}$$

The phase windings of the alternator are connected in delta. In a delta connection, the phase current is less than the line current by a factor of 1.732:

$$I_{P(Alt)} = \frac{I_{L(Alt)}}{1.732}$$

$$I_{P(Alt)} = \frac{23.095\ A}{1.732}$$

$$I_{P(Alt)} = 13.334\ A$$

The phase voltage of a delta is the same as the line voltage:

$$E_{P(Alt)} = 240\ V$$

Because the load in this example is pure resistive, the power factor has a value of unity, or 1. Power is calculated by using the line values of voltage and current:

$$P = 1.732 \times E_L \times I_L \times PF$$
$$P = 1.732 \times 240\ V \times 23.095\ A \times 1$$
$$P = 9600.13\ W$$

■ EXAMPLE 26-3

The phase windings of an alternator are connected in wye. The alternator produces a line voltage of 440 V and supplies power to two resistive loads. One load contains resistors with a value of 4 Ω each, connected in wye. The second load contains resistors with a value of 6 Ω each, connected in delta *(Figure 26–20)*. Find the following circuit values:

$E_{L(Load\ 2)}$—line voltage of Load 2

$E_{P(Load\ 2)}$—phase voltage of Load 2

$I_{P(Load\ 2)}$—phase current of Load 2

$I_{L(Load\ 2)}$—line current to Load 2

$E_{P(Load\ 1)}$—phase voltage of Load 1

$I_{P(Load\ 1)}$—phase current of Load 1

$I_{L(Load\ 1)}$—line current to Load 1

$I_{L(Alt)}$—line current delivered by the alternator

$I_{P(Alt)}$—phase current of the alternator

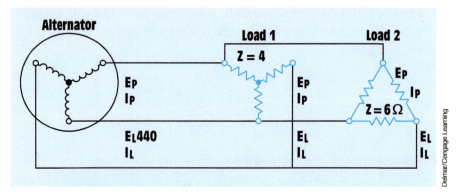

FIGURE 26–20 Calculating three-phase values using a wye-connected source and two three-phase loads (example circuit).

$E_{P(Alt)}$—phase voltage of the alternator

P—true power

Solution

Both loads are connected directly to the output of the alternator. The line voltage for both Load 1 and Load 2 is the same as the line voltage of the alternator:

$$E_{L(Load\ 2)} = 440\ V$$

$$E_{L(Load\ 1)} = 440\ V$$

Load 2 is connected as a delta. The phase voltage is the same as the line voltage:

$$E_{P(Load\ 2)} = 440\ V$$

Each of the resistors that constitutes a phase of Load 2 has a value of 6 Ω. The amount of phase current can be found using Ohm's law:

$$I_{P(Load\ 2)} = \frac{E_{P(Load\ 2)}}{Z}$$

$$I_{P(Load\ 2)} = \frac{440\ V}{6\ \Omega}$$

$$I_{P(Load\ 2)} = 73.333\ A$$

The line current supplying a delta-connected load is 1.732 times greater than the phase current. The amount of line current needed for Load 2 can be calculated by increasing the phase current value by 1.732:

$$I_{L(Load\ 2)} = I_{P(Load\ 2)} \times 1.732$$

$$I_{L(Load\ 2)} = 73.333\ A \times 1.732$$
$$I_{L(Load\ 2)} = 127.013\ A$$

The resistors of Load 1 are connected to form a wye. The phase voltage of a wye connection is less than the line voltage by a factor of 1.732:

$$E_{P(Load\ 1)} = \frac{E_{L(Load\ 1)}}{1.732}$$
$$E_{P(Load\ 1)} = \frac{440\ V}{1.732}$$
$$E_{P(Load\ 1)} = 254.042\ V$$

Now that the voltage applied to each of the 4-Ω resistors is known, the phase current can be calculated using Ohm's law:

$$I_{P(Load\ 1)} = \frac{E_{P(Load\ 1)}}{Z}$$
$$I_{P(Load\ 1)} = \frac{254.042\ V}{4\ \Omega}$$
$$I_{P(Load\ 1)} = 63.511\ A$$

The line current supplying a wye-connected load is the same as the phase current. Therefore, the amount of line current needed to supply Load 1 is

$$I_{L(Load\ 1)} = 63.511\ A$$

The alternator must supply the line current needed to operate both loads. In this example, both loads are resistive. The total line current supplied by the alternator is the sum of the line currents of the two loads:

$$I_{L(Alt)} = I_{L(Load\ 1)} + I_{L(Load\ 2)}$$
$$I_{L(Alt)} = 63.511\ A + 127.013\ A$$
$$I_{L(Alt)} = 190.524\ A$$

Because the phase windings of the alternator in this example are connected in a wye, the phase current is the same as the line current:

$$I_{P(Alt)} = 190.524\ A$$

The phase voltage of the alternator is less than the line voltage by a factor of 1.732:

$$E_{P(Alt)} = \frac{440\ V}{1.732}$$
$$E_{P(Alt)} = 254.042\ V$$

Both of the loads in this example are resistive and have a unity power factor of 1. The total power in this circuit can be found by using the line voltage and total line current supplied by the alternator:

$$P = 1.732 \times E_L \times I_L \times PF$$
$$P = 1.732 \times 440 \text{ V} \times 190.524 \text{ A} \times 1$$
$$P = 145{,}194.53 \text{ W}$$

■ EXAMPLE 26-4

A wye-connected three-phase alternator with a line voltage of 560 V supplies power to three different loads *(Figure 26–21)*. The first load is formed by three resistors with a value of 6 Ω each, connected in a wye; the second load comprises three inductors with an inductive reactance of 10 Ω each, connected in delta; and the third load comprises three capacitors with a capacitive reactance of 8 Ω each, connected in wye. Find the following circuit values:

$E_{L(Load\ 3)}$—line voltage of Load 3 (capacitive)

$E_{P(Load\ 3)}$—phase voltage of Load 3 (capacitive)

$I_{P(Load\ 3)}$—phase current of Load 3 (capacitive)

$I_{L(Load\ 3)}$—line current to Load 3 (capacitive)

$E_{L(Load\ 2)}$—line voltage of Load 2 (inductive)

$E_{P(Load\ 2)}$—phase voltage of Load 2 (inductive)

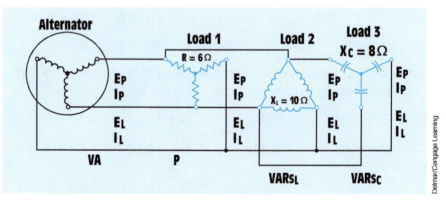

FIGURE 26–21 Calculating three-phase values with a wye-connected source supplying power to a resistive, inductive, and capacitive load (example circuit).

$I_{P(Load\ 2)}$—phase current of Load 2 (inductive)

$I_{L(Load\ 2)}$—line current to Load 2 (inductive)

$E_{L(Load\ 1)}$—line voltage of Load 1 (resistive)

$E_{P(Load\ 1)}$—phase voltage of Load 1 (resistive)

$I_{P(Load\ 1)}$—phase current of Load 1 (resistive)

$I_{L(Load\ 1)}$—line current to Load 1 (resistive)

$I_{L(Alt)}$—line current delivered by the alternator

$E_{P(Alt)}$—phase voltage of the alternator

P—true power

$VARs_L$—reactive power of the inductive load

$VARs_C$—reactive power of the capacitive load

VA—apparent power

PF—power factor

Solution

All three loads are connected to the output of the alternator. The line voltage connected to each load is the same as the line voltage of the alternator:

$$E_{L(Load\ 3)} = 560\ V$$
$$E_{L(Load\ 2)} = 560\ V$$
$$E_{L(Load\ 1)} = 560\ V$$

26-7 Load 3 Calculations

Load 3 is formed from three capacitors with a capacitive reactance of 8 ohms each, connected in a wye. Because this load is wye connected, the phase voltage is less than the line voltage by a factor of 1.732:

$$E_{P(Load\ 3)} = \frac{E_{L(Load\ 3)}}{1.732}$$

$$E_{P(Load\ 3)} = \frac{560\ V}{1.732}$$

$$E_{P(Load\ 3)} = 323.326\ V$$

Now that the voltage applied to each capacitor is known, the phase current can be calculated using Ohm's law:

$$I_{P(Load\ 3)} = \frac{E_{P(Load\ 3)}}{X_C}$$

$$I_{P(Load\ 3)} = \frac{323.326\ V}{8\ \Omega}$$

$$I_{P(Load\ 3)} = 40.416\ A$$

The line current required to supply a wye-connected load is the same as the phase current:

$$I_{L(Load\ 3)} = 40.416\ A$$

The reactive power of Load 3 can be found using a formula similar to the formula for calculating apparent power. Because Load 3 is pure capacitive, the current and voltage are 90° out of phase with each other and the power factor is zero:

$$VARs_C = 1.732 \times E_{L(Load\ 3)} \times I_{L(Load\ 3)}$$
$$VARs_C = 1.732 \times 560\ V \times 40.416\ A$$
$$VARs_C = 39,200.287$$

26–8 Load 2 Calculations

Load 2 comprises three inductors connected in a delta with an inductive reactance of 10 ohms each. Because the load is connected in delta, the phase voltage is the same as the line voltage:

$$E_{L(Load\ 2)} = 560\ V$$

The phase current can be calculated by using Ohm's law:

$$I_{P(Load\ 2)} = \frac{E_{P(Load\ 2)}}{X_L}$$

$$I_{P(Load\ 2)} = \frac{560\ V}{10\ \Omega}$$

$$I_{P(Load\ 2)} = 56\ A$$

The amount of line current needed to supply a delta-connected load is 1.732 times greater than the phase current of the load:

$$I_{L(Load\ 2)} = I_{P(Load\ 2)} \times 1.732$$
$$I_{L(Load\ 2)} = 56\ A \times 1.732$$
$$I_{L(Load\ 2)} = 96.992\ A$$

Because Load 2 is made up of inductors, the reactive power can be calculated using the line values of voltage and current supplied to the load:

$$\text{VARs}_L = 1.732 \times E_{L(\text{Load 2})} \times I_{L(\text{Load 2})}$$
$$\text{VARs}_L = 1.732 \times 560 \text{ V} \times 96.992 \text{ A}$$
$$\text{VARs}_L = 94{,}074.481$$

26–9 Load 1 Calculations

Load 1 consists of three resistors with a resistance of 6 ohms each, connected in wye. In a wye connection, the phase voltage is less than the line voltage by a factor of 1.732. The phase voltage for Load 1 is the same as the phase voltage for Load 3:

$$E_{P(\text{Load 1})} = 323.326 \text{ V}$$

The amount of phase current can now be calculated using the phase voltage and the resistance of each phase:

$$I_{P(\text{Load 1})} = \frac{E_{P(\text{Load 1})}}{R}$$
$$I_{P(\text{Load 1})} = \frac{323.326 \text{ V}}{6 \, \Omega}$$
$$I_{P(\text{Load 1})} = 53.888 \text{ A}$$

Because the resistors of Load 1 are connected in a wye, the line current is the same as the phase current:

$$I_{L(\text{Load 1})} = 53.888 \text{ A}$$

Because Load 1 is pure resistive, true power can be calculated using the line and phase current values:

$$P = 1.732 \times E_{L(\text{Load 1})} \times I_{L(\text{Load 1})}$$
$$P = 1.732 \times 560 \text{ V} = 53.888 \text{ A}$$
$$P = 52{,}267.049 \text{ W}$$

26–10 Alternator Calculations

The alternator must supply the line current for each of the loads. In this problem, however, the line currents are out of phase with each other. To find the total line current delivered by the alternator, vector addition must be used. The current flow in Load 1 is resistive and in phase with the line voltage. The current flow in Load 2 is inductive and lags the line voltage by 90°. The current flow in Load 3 is capacitive and leads the line voltage by 90°. A formula similar to the formula used to find total current flow in an RLC parallel circuit can be employed to find the total current delivered by the alternator:

$$I_{L(Alt)} = \sqrt{I_{L(Load\ 1)}^2 + (I_{L(Load\ 2)} - I_{L(Load\ 3)})^2}$$

$$I_{L(Alt)} = \sqrt{(53.888\ A)^2 + (96.992\ A - 40.416\ A)^2}$$

$$I_{L(Alt)} = 78.133\ A$$

The apparent power can now be found using the line voltage and current values of the alternator:

$$VA = 1.732 \times E_{L(Alt)} \times I_{L(Alt)}$$
$$VA = 1.732 \times 560V \times 78.133\ A$$
$$VA = 75,782.759$$

The circuit power factor is the ratio of apparent power and true power:

$$PF = \frac{W}{VA}$$

$$PF = \frac{52,267.049\ W}{75,782.759\ VA}$$

$$PF = 69\%$$

26–11 Power Factor Correction

Correcting the power factor of a three-phase circuit is similar to the procedure used to correct the power factor of a single-phase circuit.

■ EXAMPLE 26-5

A three-phase motor is connected to a 480-V, 60-Hz line *(Figure 26–22)*. A clamp-on ammeter indicates a running current of 68 A at full load, and a three-phase wattmeter indicates a true power of 40,277 W. Calculate the motor power factor

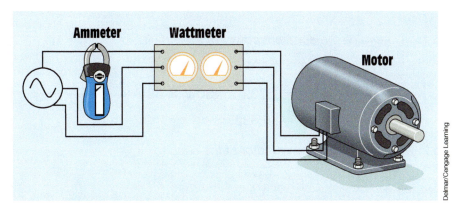

FIGURE 26–22 Determining apparent and true power for a three-phase motor.

first. Then find the amount of capacitance needed to correct the power factor to 95%. Assume that the capacitors used for power factor correction are to be connected in wye and the capacitor bank is then to be connected in parallel with the motor.

When determining the amount of capacitance needed to correct power factor, it is helpful to follow a procedure. The procedure in this example will consist of nine steps.

Step 1: Determine the apparent power of the circuit.

Step 2: Determine the power factor of the circuit.

Step 3: Determine the reactive power of the circuit.

Step 4: Determine the amount of apparent power that would produce the desired power factor.

Step 5: Determine the amount of reactive power that would produce the desired amount of apparent power.

Step 6: Determine the capacitive VARs necessary to produce the desired reactive power.

Step 7: Determine the amount of capacitive current necessary to produce the capacitive VARs needed.

Step 8: Determine the capacitive reactance of each capacitor.

Step 9: Determine the capacitance value of each capacitor.

Solution

Step 1: Determine the apparent power of the circuit.

$$VA = 1.732 \times E_L \times I_L$$

$$VA = 1.732 \times 480 \text{ V} \times 68 \text{ A}$$

$$VA = 56,532.48$$

Step 2: Determine the power factor of the circuit.

$$PF = \frac{P}{VA}$$

$$PF = \frac{40,277 \text{ W}}{56,532.48 \text{ V}}$$

$$PF = 0.7124 \text{ or } 71.24\%$$

Step 3: Determine the reactive power of the circuit.

$$VARs_L = \sqrt{VA^2 - P^2}$$

$$VARs_L = \sqrt{56,532.46^2 - 40,277^2}$$

$$VARs_L = 39,669.693$$

Step 4: Determine the amount of apparent power that would produce the desired power factor.

$$VA = \frac{P}{PF}$$

$$VA = \frac{40,277}{0.95}$$

$$VA = 42,396.842$$

Step 5: Determine the amount of reactive power that would produce the desired amount of apparent power.

$$VARs_L = \sqrt{VA^2 - P^2}$$

$$VARs_L = \sqrt{42,396.842^2 - 40,277^2}$$

$$VARs_L = 13,238.409$$

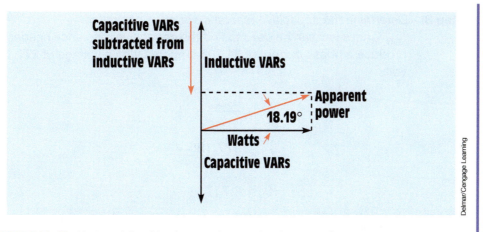

FIGURE 26-23 Vector relationship of powers to correct motor power factor.

Step 6: *Determine the capacitive VARs necessary to produce the desired reactive power.*

$$VARs_C = VARs_{present\ time} - VARs_{desired}$$

$$VARs_C = 39{,}669.693 - 13{,}238.409$$

$$VARs_C = 26{,}431.284$$

To correct the power factor to 95%, the inductive VARs must be reduced from 39,669.693 to 13,238.409. This can be done by connecting a bank of capacitors in the circuit that will produce a total of 26,431.284 capacitive VARs. This amount of capacitive VARs will reduce the inductive VARs to the desired amount *(Figure 26–23)*.

Step 7: *Determine the amount of capacitive current necessary to produce the capacitive VARs needed.*

$$I_C = \frac{VARs_C}{E_L \times 1.732}$$

$$I_C = \frac{26{,}431.284\ VARs_C}{480 \times 1.732}$$

$$I_C = 31.793\ A$$

The capacitive load bank is to be connected in a wye. Therefore, the phase current is the same as the line current. The phase voltage, however, is less than the line voltage by a factor of 1.732. The phase voltage is 277.136 volts.

Step 8: *Determine the capacitive reactance of each capacitor.*

Ohm's law can be used to find the capacitive reactance needed to produce a phase current of 31.793 amperes with a voltage of 277.136 volts:

$$X_C = \frac{E_{Phase}}{I_{Phase}}$$

$$X_C = \frac{277.136 \text{ V}}{31.793 \text{ A}}$$

$$X_C = 8.717 \text{ }\Omega$$

Step 9: *Determine the capacitance value of each capacitor:*

$$C = \frac{1}{2\pi f X_C}$$

$$C = \frac{1}{377 \times 8.717}$$

$$C = 304.293 \text{ }\mu\text{F}$$

When a bank of wye-connected capacitors with a value of 304.293 μF is each connected in parallel with the motor, the power factor is corrected to 95% *(Figure 26–24)*.

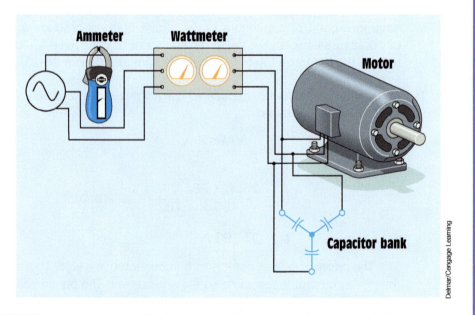

FIGURE 26–24 A wye-connected bank of capacitors is used to correct motor power factor.

Summary

- The voltages of a three-phase system are 120° out of phase with each other.
- The two types of three-phase connections are wye and delta.
- Wye connections are characterized by the fact that one terminal of each of the devices is connected together.
- In a wye connection, the phase voltage is less than the line voltage by a factor of 1.732. The phase current and line current are the same.
- In a delta connection, the phase voltage is the same as the line voltage. The phase current is less than the line current by a factor of 1.732.

Review Questions

1. How many degrees out of phase with each other are the voltages of a three-phase system?
2. What are the two main types of three-phase connections?
3. A wye-connected load has a voltage of 480 V applied to it. What is the voltage dropped across each phase?
4. A wye-connected load has a phase current of 25 A. How much current is flowing through the lines supplying the load?
5. A delta connection has a voltage of 560 V connected to it. How much voltage is dropped across each phase?
6. A delta connection has 30 A of current flowing through each phase winding. How much current is flowing through each of the lines supplying power to the load?
7. A three-phase load has a phase voltage of 240 V and a phase current of 18 A. What is the apparent power of this load?
8. If the load in Question 7 is connected in a wye, what would be the line voltage and line current supplying the load?
9. An alternator with a line voltage of 2400 V supplies a delta-connected load. The line current supplied to the load is 40 A. Assuming the load is a balanced three-phase load, what is the impedance of each phase?
10. What is the apparent power of the circuit in Question 9?

Practical Applications

You are working in an industrial plant. A bank of capacitors is to be used to correct the power factor of a 480-V, three-phase motor. The capacitor bank is connected in a wye. Each of the three capacitors is rated at 25 µF and 600 VDC. You have been told to reconnect these capacitors in delta. Can these capacitors be changed from wye to delta without harm to the capacitors? ■

Practical Applications

You are a journeyman electrician working in an industrial plant. A 480-V, three-phase, 60-Hz power panel has a current draw of 216 A. A three-phase wattmeter indicates a true power of 86 kW. You have been instructed to reduce the current draw on the panel by adding three 400-µF capacitors to the system. The capacitors are to be connected in wye. Each capacitor will be connected to one of the three-phase lines. What should be the current flow on the system after the bank of capacitors is added to the circuit? ■

Practical Applications

You are an electrician working in an industrial plant. A 30-hp three-phase induction motor has a current draw of 36 amperes at full load. The motor is connected to a 480-volt line. A three-phase wattmeter indicates a true power of 22 kW. Determine the power factor of the motor and the amount of capacitance needed to correct the power factor to 95%. Also determine the minimum voltage rating of the capacitors. The capacitors are to be connected in wye. ■

Practice Problems

1. Refer to the circuit shown in *Figure 26–18* to answer the following questions, but assume that the alternator has a line voltage of 240 V and the load has an impedance of 12 Ω per phase. Find all the missing values.

   $E_{P(A)}$ _____ $E_{P(L)}$ _____

$I_{P(A)}$ _____ $I_{P(L)}$ _____

$E_{L(A)}$ 240 V $E_{L(L)}$ _____

$I_{L(A)}$ _____ $I_{L(L)}$ _____

P _____ $Z_{(PHASE)}$ 12 Ω

2. Refer to the circuit shown in *Figure 26–19* to answer the following questions, but assume that the alternator has a line voltage of 4160 V and the load has a resistance of 60 Ω per phase. Find all the missing values.

$E_{P(A)}$ _____ $E_{P(L)}$ _____

$I_{P(A)}$ _____ $I_{P(L)}$ _____

$E_{L(A)}$ 4160 V $E_{L(L)}$ _____

$I_{L(A)}$ _____ $I_{L(L)}$ _____

P _____ $Z_{(PHASE)}$ 60 Ω

3. Refer to the circuit shown in *Figure 26–20* to answer the following questions, but assume that the alternator has a line voltage of 560 V. Load 1 has an resistance of 5 Ω per phase, and Load 2 has a resistance of 8 Ω per phase. Find all the missing values.

$E_{P(A)}$ _____ $E_{P(L1)}$ _____ $E_{P(L2)}$ _____

$I_{P(A)}$ _____ $I_{P(L1)}$ _____ $I_{P(L2)}$ _____

$E_{L(A)}$ 560 V $E_{L(L1)}$ _____ $E_{L(L2)}$ _____

$I_{L(A)}$ _____ $I_{L(L1)}$ _____ $I_{L(L2)}$ _____

P _____ $Z_{(PHASE)}$ 5 Ω $Z_{(PHASE)}$ 8 Ω

4. Refer to the circuit shown in *Figure 26–21* to answer the following questions, but assume that the alternator has a line voltage of 480 V. Load 1 has a resistance of 12 Ω per phase. Load 2 has an inductive reactance of 16 Ω per phase, and Load 3 has a capacitive reactance of 10 Ω per phase. Find all the missing values.

$E_{P(A)}$ _____ $E_{P(L1)}$ _____ $E_{P(L2)}$ _____ $E_{P(L3)}$ _____

$I_{P(A)}$ _____ $I_{P(L1)}$ _____ $I_{P(L2)}$ _____ $I_{P(L3)}$ _____

$E_{L(A)}$ 480 V $E_{L(L1)}$ _____ $E_{L(L2)}$ _____ $E_{L(L3)}$ _____

$I_{L(A)}$ _____ $I_{L(L1)}$ _____ $I_{L(L2)}$ _____ $I_{L(L3)}$ _____

VA _____ $R_{(PHASE)}$ 12 Ω $X_{L(PHASE)}$ 16 Ω $X_{C(PHASE)}$ 10 Ω

                        P _____ $VARS_L$ _____ $VARS_C$ _____

XIII Transformers

Unit 27
Single-Phase Transformers

OUTLINE

- **27–1** Single-Phase Transformers
- **27–2** Isolation Transformers
- **27–3** Autotransformers
- **27–4** Transformer Polarities
- **27–5** Voltage and Current Relationships in a Transformer
- **27–6** Testing the Transformer
- **27–7** Transformer Nameplates
- **27–8** Determining Maximum Current
- **27–9** Transformer Impedance

KEY TERMS

- Autotransformers
- Constant-current transformer
- Control transformer
- Current regulator
- Distribution transformer
- Excitation current
- Flux leakage
- Inrush current
- Isolation transformers
- Laminated
- Neutral conductor
- Primary winding
- Secondary winding
- Step-down transformer
- Step-up transformer
- Tape-wound core
- Toroid core
- Transformer
- Turns ratio
- Volts-per-turn ratio

Why You Need to Know

Understanding how transformers change values of voltage and current is important to the information presented in later units. Many AC motors, for example, operate on these principles. This unit

- discusses transformers and how they are divided into three major types.
- determines values of voltage and current and illustrates different methods that can be employed to determine these values.
- describes how windings determine the primary and secondary voltage and how the nameplate rating provides key information when providing protection.
- discusses installation and testing.

Objectives

After studying this unit, you should be able to

- discuss the different types of transformers.
- calculate values of voltage, current, and turns for single-phase transformers using formulas.
- calculate values of voltage, current, and turns for single-phase transformers using the turns ratio.
- connect a transformer and test the voltage output of different windings.
- discuss polarity markings on a schematic diagram.
- test a transformer to determine the proper polarity marks.

Preview

Transformers are among the most common devices found in the electrical field. They range in size from less than one cubic inch to the size of rail cars. Their ratings can range from milli-volt-amperes (mVA) to giga-volt-amperes (GVA). It is imperative that anyone working in the electrical field have an understanding of transformer types and connections. This unit presents transformers intended for use in single-phase installations. The two main types of voltage transformers, isolation transformers and autotransformers, are discussed. ■

27-1 Single-Phase Transformers

A **transformer** is a magnetically operated machine that can change values of voltage, current, and impedance without a change of frequency. Transformers are the most efficient machines known. Their efficiencies commonly range from 90% to 99% at full load. Transformers can be divided into three classifications:

1. Isolation transformer
2. Autotransformer
3. Current transformer (current transformers were discussed in Unit 25)

All values of a transformer are proportional to its turns ratio. This does not mean that the exact number of turns of wire on each winding must be known to determine different values of voltage and current for a transformer. What must be known is the *ratio* of turns. For example, assume a transformer has two windings. One winding, the primary, has 1000 turns of wire; and the other, the secondary, has 250 turns of wire *(Figure 27–1)*. The **turns ratio** of this transformer is 4 to 1, or 4:1 (1000 turns/250 turns = 4). This indicates

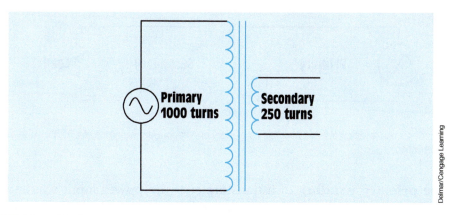

FIGURE 27–1 All values of a transformer are proportional to its turns ratio.

there are four turns of wire on the primary for every one turn of wire on the secondary.

Transformer Formulas

Different formulas can be used to find the values of voltage and current for a transformer. The following is a list of standard formulas, where

N_P = number of turns in the primary
N_S = number of turns in the secondary
E_P = voltage of the primary
E_S = voltage of the secondary
I_P = current in the primary
I_S = current in the secondary

$$\frac{E_P}{E_S} = \frac{N_P}{N_S}$$

$$\frac{E_P}{E_S} = \frac{I_S}{I_P}$$

$$\frac{N_P}{N_S} = \frac{I_S}{I_P}$$

or

$$E_P \times N_S = E_S \times N_P$$

$$E_P \times I_P = E_S \times I_S$$

$$N_P \times I_P = N_S \times I_S$$

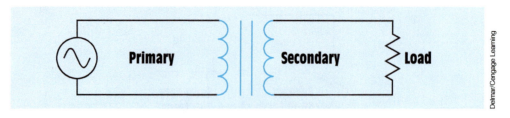

FIGURE 27–2 An isolation transformer has its primary and secondary windings electrically separated from each other.

The **primary winding** of a transformer is the power input winding. It is the winding that is connected to the incoming power supply. The **secondary winding** is the load winding, or output winding. It is the side of the transformer that is connected to the driven load *(Figure 27–2)*.

27–2 Isolation Transformers

The transformers shown in *Figure 27–1* and *Figure 27–2* are **isolation transformers.** This means that the secondary winding is physically and electrically isolated from the primary winding. There is no electric connection between the primary and secondary winding. This transformer is magnetically coupled, not electrically coupled. This line isolation is often a very desirable characteristic. The isolation transformer greatly reduces any voltage spikes that originate on the supply side before they are transferred to the load side. Some isolation transformers are built with a turns ratio of 1:1. A transformer of this type has the same input and output voltages and is used for the purpose of isolation only.

The reason that the isolation transformer can greatly reduce any voltage spikes before they reach the secondary is because of the rise time of current through an inductor. Recall from Unit 11 that DC in an inductor rises at an exponential rate *(Figure 27–3)*. As the current increases in value, the expanding magnetic field cuts through the conductors of the coil and induces a voltage that is opposed to the applied voltage. The amount of induced voltage is proportional to the rate of change of current. This simply means that the faster current attempts to increase, the greater the opposition to that increase is. Spike voltages and currents are generally of very short duration, which means that they increase in value very rapidly *(Figure 27–4)*. This rapid change of value causes the opposition to the change to increase just as rapidly. By the time the spike has been transferred to the secondary winding of the transformer, it has been eliminated or greatly reduced *(Figure 27–5)*.

The basic construction of an isolation transformer is shown in *Figure 27–6*. A metal core is used to provide good magnetic coupling between the two windings. The core is generally made of laminations stacked together. Laminating the core helps reduce power losses caused by eddy current induction.

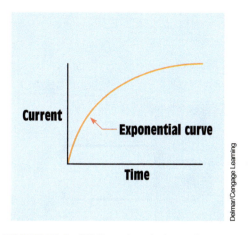

FIGURE 27–3 DC through an inductor rises at an exponential rate.

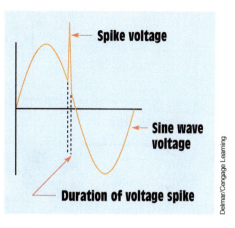

FIGURE 27–4 Voltage spikes are generally of very short duration.

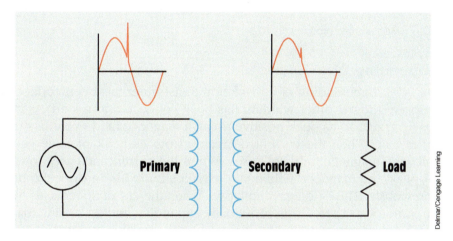

FIGURE 27–5 The isolation transformer greatly reduces the voltage spike.

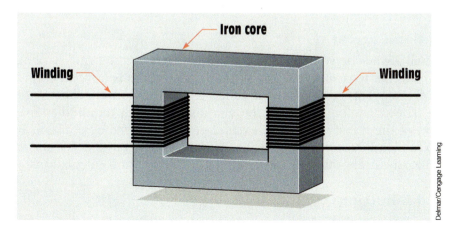

FIGURE 27–6 Basic construction of an isolation transformer.

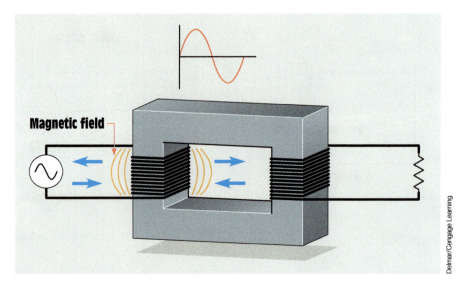

FIGURE 27–7 Magnetic field produced by AC.

Basic Operating Principles

In *Figure 27–7,* one winding of an isolation transformer has been connected to an AC supply, and the other winding has been connected to a load. As current increases from zero to its peak positive point, a magnetic field expands outward around the coil. When the current decreases from its peak positive point toward zero, the magnetic field collapses. When the current increases toward its negative peak, the magnetic field again expands but with an opposite polarity of that previously. The field again collapses when the current decreases from its negative peak toward zero. This continually expanding and collapsing magnetic field cuts the windings of the primary and induces a voltage into it. This induced voltage opposes the applied voltage and limits the current flow of the primary. When a coil induces a voltage into itself, it is known as *self-induction*.

Excitation Current

There will always be some amount of current flow in the primary of any voltage transformer regardless of type or size even if there is no load connected to the secondary. This current flow is called the **excitation current** of the transformer. The excitation current is the amount of current required to magnetize the core of the transformer. The excitation current remains constant from no load to full load. As a general rule, the excitation current is such a small part of the full-load current that it is often omitted when making calculations.

Mutual Induction

Because the secondary windings of an isolation transformer are wound on the same core as the primary, the magnetic field produced by the primary winding

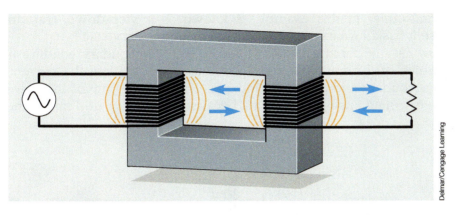

FIGURE 27–8 The magnetic field of the primary induces a voltage into the secondary.

also cuts the windings of the secondary *(Figure 27–8)*. This continually changing magnetic field induces a voltage into the secondary winding. The ability of one coil to induce a voltage into another coil is called *mutual induction*. The amount of voltage induced in the secondary is determined by the ratio of the number of turns of wire in the secondary to those in the primary. For example, assume the primary has 240 turns of wire and is connected to 120 VAC. This gives the transformer a **volts-per-turn ratio** of 0.5 (120 V/240 turns = 0.5 V per turn). Now assume the secondary winding contains 100 turns of wire. Because the transformer has a volts-per-turn ratio of 0.5 volt per turn, the secondary voltage is 50 volts (100 turns × 0.5 V/turn = 50 V per turn).

Transformer Calculations

In the following examples, values of voltage, current, and turns for different transformers are calculated.

Assume that the isolation transformer shown in *Figure 27–2* has 240 turns of wire on the primary and 60 turns of wire on the secondary. This is a ratio of 4:1 (240 turns/60 turns = 4). Now assume that 120 volts are connected to the primary winding. What is the voltage of the secondary winding?

$$\frac{E_P}{E_S} = \frac{N_P}{N_S}$$

$$\frac{120\ V}{E_S} = \frac{240\ \text{turns}}{60\ \text{turns}}$$

$$240\ \text{turns}\ E_S = 7200\ \text{V-turns}$$

$$E_S = \frac{7200\ \text{V-turns}}{240\ \text{turns}}$$

$$E_S = 30\ V$$

The transformer in this example is known as a **step-down transformer** because it has a lower secondary voltage than primary voltage.

Now assume that the load connected to the secondary winding has an impedance of 5 ohms. The next problem is to calculate the current flow in the secondary and primary windings. The current flow of the secondary can be calculated using Ohm's law because the voltage and impedance are known:

$$I = \frac{E}{Z}$$

$$I = \frac{30 \text{ V}}{5 \text{ }\Omega}$$

$$I = 6 \text{ A}$$

Now that the amount of current flow in the secondary is known, the primary current can be calculated using the formula

$$\frac{E_P}{E_S} = \frac{I_S}{I_P}$$

$$\frac{120 \text{ V}}{30 \text{ V}} = \frac{6 \text{ A}}{I_P}$$

$$120 \text{ V } I_P = 180 \text{ VA}$$

$$I_P = \frac{180 \text{ VA}}{120 \text{ V}}$$

$$I_P = 1.5 \text{ A}$$

Notice that the primary voltage is higher than the secondary voltage but the primary current is much less than the secondary current. *A good rule for any type of transformer is that power in must equal power out.* If the primary voltage and current are multiplied together, the product should equal the product of the voltage and current of the secondary:

Primary	Secondary
120 × 1.5 = 180 VA	30 × 6 = 180 VA

In this example, assume that the primary winding contains 240 turns of wire and the secondary contains 1200 turns of wire. This is a turns ratio of 1:5 (1200 turns/240 turns = 5). Now assume that 120 volts are connected to the primary winding. Calculate the voltage output of the secondary winding:

$$\frac{E_P}{E_S} = \frac{N_P}{N_S}$$

$$\frac{120 \text{ V}}{E_S} = \frac{240 \text{ turns}}{1200 \text{ turns}}$$

$$240 \text{ turns } E_S = 144{,}000 \text{ V-turns}$$

$$E_S = \frac{144{,}000 \text{ V-turns}}{240 \text{ turns}}$$

$$E_S = 600 \text{ V}$$

Notice that the secondary voltage of this transformer is higher than the primary voltage. This type of transformer is known as a **step-up transformer.**

Now assume that the load connected to the secondary has an impedance of 2400 ohms. Find the amount of current flow in the primary and secondary windings. The current flow in the secondary winding can be calculated using Ohm's law:

$$I = \frac{E}{Z}$$

$$I = \frac{600 \text{ V}}{2400 \text{ }\Omega}$$

$$I = 0.25 \text{ A}$$

Now that the amount of current flow in the secondary is known, the primary current can be calculated using the formula

$$\frac{E_P}{E_S} = \frac{I_S}{I_P}$$

$$\frac{120 \text{ V}}{600 \text{ V}} = \frac{0.25 \text{ A}}{I_P}$$

$$120 \text{ V } I_P = 150 \text{ VA}$$

$$I_P = 1.25 \text{ A}$$

Notice that the amount of power input equals the amount of power output:

Primary	Secondary
120 V × 1.25 A = 150 VA	600 V × 0.25 A = 150 VA

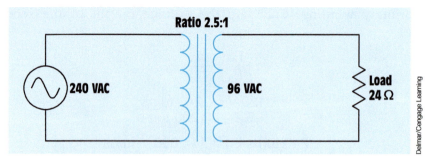

FIGURE 27–9 Calculating transformer values using the turns ratio.

Calculating Isolation Transformer Values Using the Turns Ratio

As illustrated in the previous examples, transformer values of voltage, current, and turns can be calculated using formulas. It is also possible to calculate these same values using the turns ratio. To make calculations using the turns ratio, a ratio is established that compares some number to 1, or 1 to some number. For example, assume a transformer has a primary rated at 240 volts and a secondary rated at 96 volts *(Figure 27–9)*. The turns ratio can be calculated by dividing the higher voltage by the lower voltage:

$$\text{Ratio} = \frac{240 \text{ V}}{96 \text{ V}}$$

$$\text{Ratio} = 2.5:1$$

This ratio indicates that there are 2.5 turns of wire in the primary winding for every 1 turn of wire in the secondary. The side of the transformer with the lowest voltage will always have the lowest number (1) of the ratio.

Now assume that a resistance of 24 ohms is connected to the secondary winding. The amount of secondary current can be found using Ohm's law:

$$I_s = \frac{96}{24}$$

$$I_s = 4 \text{ A}$$

The primary current can be found using the turns ratio. Recall that the volt-amperes of the primary must equal the volt-amperes of the secondary. Because the primary voltage is greater, the primary current will have to be less than the secondary current:

$$I_p = \frac{I_s}{\text{turns ratio}}$$

$$I_p = \frac{4 \text{ A}}{2.5}$$

$$I_p = 1.6 \text{ A}$$

To check the answer, find the volt-amperes of the primary and secondary:

Primary	Secondary
240 V × 1.6 A = 384 VA	96 V × 4 A = 384 VA

Now assume that the secondary winding contains 150 turns of wire. The primary turns can be found by using the turns ratio also. Because the primary voltage is higher than the secondary voltage, the primary must have more turns of wire:

$$N_P = N_S \times \text{turns ratio}$$
$$N_P = 150 \text{ turns} \times 2.5$$
$$N_P = 375 \text{ turns}$$

In the next example, assume an isolation transformer has a primary voltage of 120 volts and a secondary voltage of 500 volts. The secondary has a load impedance of 1200 ohms. The secondary contains 800 turns of wire *(Figure 27–10)*.

The turns ratio can be found by dividing the higher voltage by the lower voltage:

$$\text{Ratio} = \frac{500 \text{ V}}{120 \text{ V}}$$

$$\text{Ratio} = 1.4:17$$

The secondary current can be found using Ohm's law:

$$I_S = \frac{500 \text{ V}}{1200 \, \Omega}$$

$$I_S = 0.417 \text{ A}$$

In this example, the primary voltage is lower than the secondary voltage. Therefore, the primary current must be higher:

$$I_P = I_S \times \text{turns ratio}$$
$$I_P = 0.417 \text{ A} \times 4.17$$
$$I_P = 1.739 \text{ A}$$

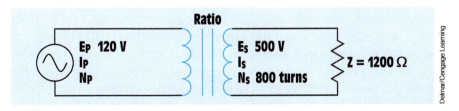

FIGURE 27–10 Calculating transformer values.

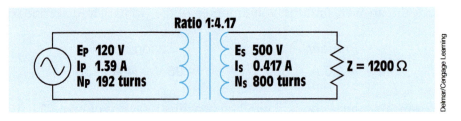

FIGURE 27–11 Transformer with calculated values.

To check this answer, calculate the volt-amperes of both windings:

Primary   Secondary
120 V × 1.739 A = 208.68 VA   500 V × 0.417 A = 208.5 VA

The slight difference in answers is caused by rounding off values.

Because the primary voltage is less than the secondary voltage, the turns of wire in the primary is less also:

$$N_P = \frac{N_S}{\text{turns ratio}}$$

$$N_P = \frac{800 \text{ turns}}{4.17}$$

$$N_P = 192 \text{ turns}$$

Figure 27–11 shows the transformer with all calculated values.

Multiple-Tapped Windings

It is not uncommon for isolation transformers to be designed with windings that have more than one set of lead wires connected to the primary or secondary. These are called *multiple-tapped windings*. The transformer shown in *Figure 27–12* contains a secondary winding rated at 24 volts. The primary winding contains several taps, however. One of the primary lead wires is labeled C and is the common for the other leads. The other leads are labeled 120 volts, 208 volts, and 240 volts. This transformer is designed in such a manner that it can be connected to different primary voltages without changing the value of the secondary voltage. In this example, it is assumed that the secondary winding has a total of 120 turns of wire. To maintain the proper turns ratio, the primary would have 600 turns of wire between C and 120 volts, 1040 turns between C and 208 volts, and 1200 turns between C and 240 volts.

The isolation transformer shown in *Figure 27–13* contains a single primary winding. The secondary winding, however, has been tapped at several points. One of the secondary lead wires is labeled C and is common to the other lead wires. When rated voltage is applied to the primary, voltages of 12 volts, 24 volts,

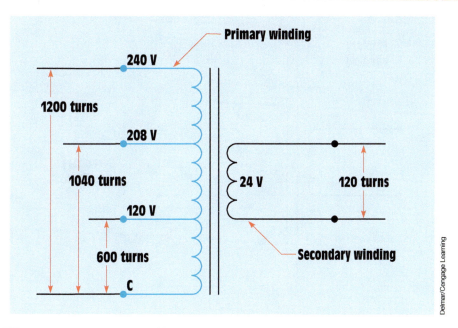

FIGURE 27–12 Transformer with multiple-tapped primary winding.

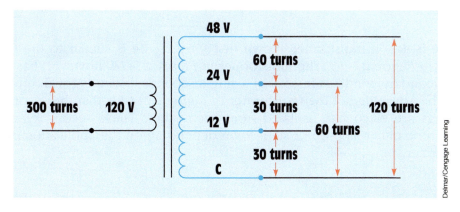

FIGURE 27–13 Transformer secondary with multiple taps.

and 48 volts can be obtained at the secondary. It should also be noted that this arrangement of taps permits the transformer to be used as a center-tapped transformer for two of the voltages. If a load is placed across the lead wires labeled C and 24, the lead wire labeled 12 volts becomes a center tap. If a load is placed across the C and 48 lead wires, the 24 volts lead wire becomes a center tap.

In this example, it is assumed that the primary winding has 300 turns of wire. To produce the proper turns ratio would require 30 turns of wire between C and 12 volts, 60 turns of wire between C and 24 volts, and 120 turns of wire between C and 48 volts.

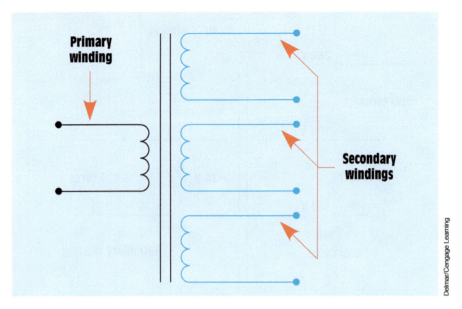

FIGURE 27–14 Transformer with multiple secondary windings.

The isolation transformer shown in *Figure 27–14* is similar to the transformer in *Figure 27–13*. The transformer in *Figure 27–14*, however, has multiple secondary windings instead of a single secondary winding with multiple taps. The advantage of the transformer in *Figure 27–14* is that the secondary windings are electrically isolated from each other. These secondary windings can be either step-up or step-down depending on the application of the transformer.

Calculating Values for Isolation Transformers with Multiple Secondaries

When calculating the values of an isolation transformer with multiple secondary windings, each secondary must be treated as a different transformer. For example, the transformer in *Figure 27–15* contains one primary winding and three secondary windings. The primary is connected to 120 VAC and contains 300 turns of wire. One secondary has an output voltage of 560 volts and a load impedance of 1000 ohms; the second secondary has an output voltage of 208 volts and a load impedance of 400 ohms; and the third secondary has an output voltage of 24 volts and a load impedance of 6 ohms. The current, turns of wire, and ratio for each secondary and the current of the primary will be found.

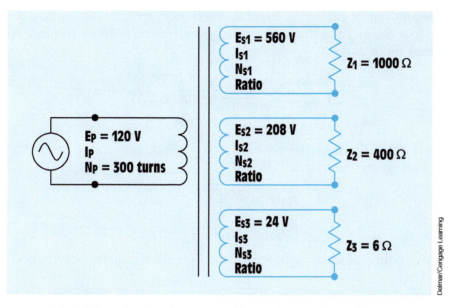

FIGURE 27–15 Calculating values for a transformer with multiple secondary windings.

The first step is to calculate the turns ratio of the first secondary. The turns ratio can be found by dividing the smaller voltage into the larger:

$$\text{Ratio} = \frac{E_{S1}}{E_P}$$

$$\text{Ratio} = \frac{560 \text{ V}}{120 \text{ V}}$$

$$\text{Ratio} = 1:4.67$$

The current flow in the first secondary can be calculated using Ohm's law:

$$I_{S1} = \frac{560 \text{ V}}{1000 \text{ V}}$$

$$I_{S1} = 0.56 \text{ A}$$

The number of turns of wire in the first secondary winding is found using the turns ratio. Because this secondary has a higher voltage than the primary, it must have more turns of wire:

$$N_{S1} = N_P \times \text{turns ratio}$$

$$N_{S1} = 300 \text{ turns} \times 4.67$$

$$N_{S1} = 1401 \text{ turns}$$

The amount of primary current needed to supply this secondary winding can be found using the turns ratio also. Because the primary has less voltage, it requires more current:

$$I_{P(FIRST\ SECONDARY)} = I_{S1} \times \text{turns ratio}$$
$$I_{P(FIRST\ SECONDARY)} = 0.56A \times 4.67$$
$$I_{P(FIRST\ SECONDARY)} = 2.61\ A$$

The turns ratio of the second secondary winding is found by dividing the higher voltage by the lower:

$$\text{Ratio} = \frac{208\ V}{120\ V}$$
$$\text{Ratio} = 1:1.73$$

The amount of current flow in this secondary can be determined using Ohm's law:

$$I_{S2} = \frac{208\ V}{400\ \Omega}$$
$$I_{S2} = 0.52\ A$$

Because the voltage of this secondary is greater than the primary, it has more turns of wire than the primary. The number of turns of this secondary is found using the turns ratio:

$$N_{S2} = N_P \times \text{turns ratio}$$
$$N_{S2} = 300\ \text{turns} \times 1.73$$
$$N_{S2} = 519\ \text{turns}$$

The voltage of the primary is less than this secondary. The primary therefore requires a greater amount of current. The amount of current required to operate this secondary is calculated using the turns ratio:

$$I_{P(SECOND\ SECONDARY)} = I_{S2} \times \text{turns ratio}$$
$$I_{P(SECOND\ SECONDARY)} = 0.52A \times 1.732$$
$$I_{P(SECOND\ SECONDARY)} = 0.9\ A$$

The turns ratio of the third secondary winding is calculated in the same way as the other two. The larger voltage is divided by the smaller:

$$\text{Ratio} = \frac{120\ V}{24\ V}$$
$$\text{Ratio} = 5:1$$

The primary current is found using Ohm's law:

$$I_{S3} = \frac{24\text{ V}}{6\text{ }\Omega}$$

$$I_{S3} = 4\text{ A}$$

The output voltage of the third secondary is less than the primary. The number of turns of wire is therefore less than the primary turns:

$$N_{S3} = \frac{N_P}{\text{turns ratio}}$$

$$N_{S3} = \frac{300\text{ turns}}{5}$$

$$N_{S3} = 60\text{ turns}$$

The primary has a higher voltage than this secondary. The primary current is therefore less by the amount of the turns ratio:

$$I_{P\text{ (THIRD SECONDARY)}} = \frac{I_{S3}}{\text{turns ratio}}$$

$$I_{P\text{ (THIRD SECONDARY)}} = \frac{4\text{ A}}{5}$$

$$I_{P\text{ (THIRD SECONDARY)}} = 0.8\text{ A}$$

The primary must supply current to each of the three secondary windings. Therefore, the total amount of primary current is the sum of the currents required to supply each secondary:

$$I_{P(TOTAL)} = I_{P1} + I_{P2} + I_{P3}$$

$$I_{P(TOTAL)} = 2.61\text{A} + 0.9\text{A} + 0.8\text{A}$$

$$I_{P(TOTAL)} = 4.31\text{ A}$$

The transformer with all calculated values is shown in *Figure 27–16*.

Distribution Transformers

A common type of isolation transformer is the **distribution transformer** *(Figure 27–17)*. This type of transformer changes the high voltage of power company distribution lines to the common 240/120 volts used to supply power to most homes and many businesses. In this example, it is assumed that the primary is connected to a 7200-volt line. The secondary is 240 volts with a center tap. The center tap is grounded and becomes the **neutral conductor** or common conductor. If voltage is measured across the entire secondary, a voltage of 240 volts is seen. If voltage is measured from either line to the center tap, half of the secondary voltage, or 120 volts, is seen *(Figure 27–18)*. The

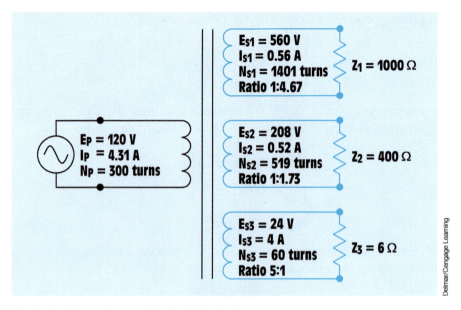

FIGURE 27–16 The transformer with all calculated values.

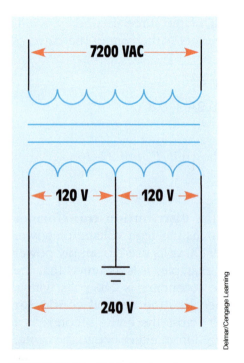

FIGURE 27–17 Distribution transformer.

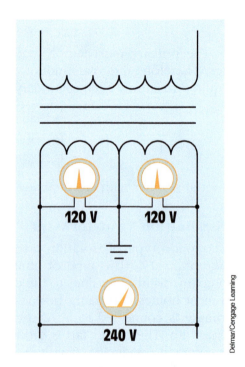

FIGURE 27–18 The voltage from either line to neutral is 120 volts. The voltage across the entire secondary winding is 240 volts.

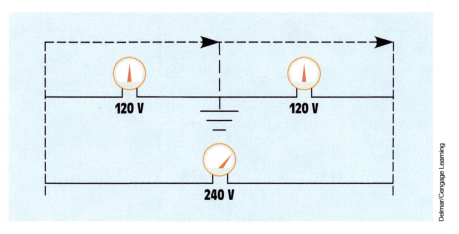

FIGURE 27–19 The voltages across the secondary are out of phase with each other.

reason this occurs is that the grounded neutral conductor becomes the center point of two out-of-phase voltages. If a vector diagram is drawn to illustrate this condition, you will see that the grounded neutral conductor is connected to the center point of the two out-of-phase voltages *(Figure 27–19)*. Loads that are intended to operate on 240 volts, such as water heaters, electric-resistance heating units, and central air conditioners are connected directly across the lines of the secondary *(Figure 27–20)*.

Loads that are intended to operate on 120 volts connect from the center tap, or neutral, to one of the secondary lines. The function of the neutral is to carry the difference in current between the two secondary lines and maintain

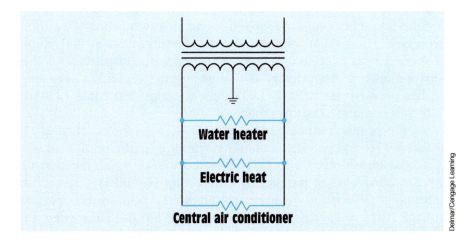

FIGURE 27–20 Loads of 240 volts connect directly across the secondary winding.

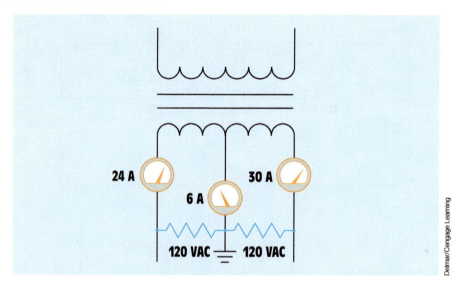

FIGURE 27–21 The neutral carries the sum of the unbalanced load.

a balanced voltage. In *Figure 27–21*, one of the secondary lines has a current flow of 30 amperes and the other has a current flow of 24 amperes. The neutral conducts the sum of the unbalanced load. In this example, the neutral current is 6 amperes (30A − 24A = 6A).

Control Transformers

Another common type of isolation transformer found throughout industry is the **control transformer** *(Figure 27–22)*. The control transformer is used to reduce the line voltage to the value needed to operate control circuits. The most common type of control transformer contains two primary windings and one secondary. The primary windings are generally rated at 240 volts each, and the secondary is rated at 120 volts. This arrangement provides a 2:1 turns ratio between each of the primary windings and the secondary. For example, assume that each of the primary windings contains 200 turns of wire. The secondary will contain 100 turns of wire.

One of the primary windings in *Figure 27–23* is labeled H_1 and H_2. The other is labeled H_3 and H_4. The secondary winding is labeled X_1 and X_2. If the primary of the transformer is to be connected to 240 volts, the two primary windings are connected in parallel by connecting H_1 and H_3 together and H_2 and H_4 together. When the primary windings are connected in parallel, the same voltage is applied across both windings. This has the same effect as using one primary winding with a total of 200 turns of wire. A turns ratio of 2:1 is maintained, and the secondary voltage is 120 volts.

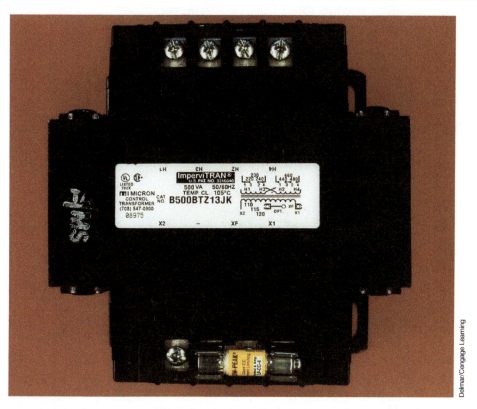

FIGURE 27–22 Control transformer with fuse protection added to the secondary winding.

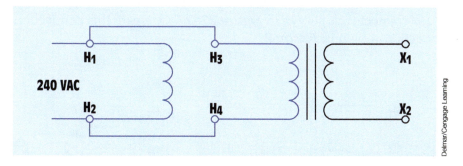

FIGURE 27–23 Control transformer connected for 240-volt operation.

If the transformer is to be connected to 480 volts, the two primary windings are connected in ser7ies by connecting H$_2$ and H$_3$ together *(Figure 27–24)*. The incoming power is connected to H$_1$ and H$_4$. Series-connecting the primary windings has the effect of increasing the number of turns in the primary to 400. This produces a turns ratio of 4:1. When 480 volts are connected to the primary, the secondary voltage will remain at 120.

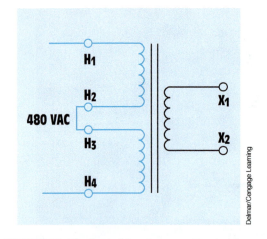

FIGURE 27–24 Control transformer connected for 480-volt operation.

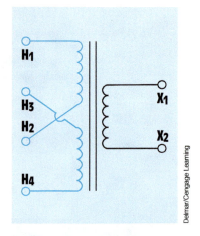

FIGURE 27–25 The primary windings of a control transformer are crossed.

The primary leads of a control transformer are generally cross-connected (*Figure 27–25*). This is done so that metal links can be used to connect the primary for 240- or 480-volt operation. If the primary is to be connected for 240-volt operation, the metal links will be connected under screws (*Figure 27–26*). Notice that leads H_1 and H_3 are connected together and leads H_2 and H_4 are connected together. Compare this connection with the connection shown in *Figure 27–23*.

If the transformer is to be connected for 480-volt operation, terminals H_2 and H_3 are connected as shown in *Figure 27–27*. Compare this connection with the connection shown in *Figure 27–24*.

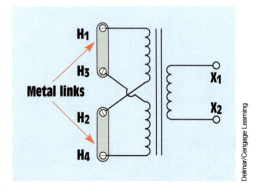

FIGURE 27–26 Metal links connect transformer for 240-volt operation.

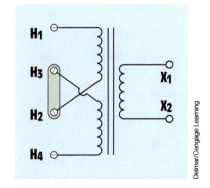

FIGURE 27–27 Control transformer connected for 480-volt operation.

Transformer Core Types

Several types of cores are used in the construction of transformers. Most cores are made from thin steel punchings **laminated** together to form a solid metal core. The core for a 600-mega-volt-ampere (MVA) three-phase transformer is shown in *Figure 27–28*. Laminated cores are preferred because a thin layer of oxide forms on the surface of each lamination and acts as an insulator to reduce the formation of eddy currents inside the core material. The amount of core material needed for a particular transformer is determined by the power rating of the transformer. The amount of core material must be sufficient to prevent saturation at full load. The type and shape of the core generally determine the amount of magnetic coupling between the windings and to some extent the efficiency of the transformer.

The transformer illustrated in *Figure 27–29* is known as a core-type transformer. The windings are placed around each end of the core material. As a general rule, the low-voltage winding is placed closest to the core and the high-voltage winding is placed over the low-voltage winding.

FIGURE 27–28 Core of a 600-MVA three-phase transformer.

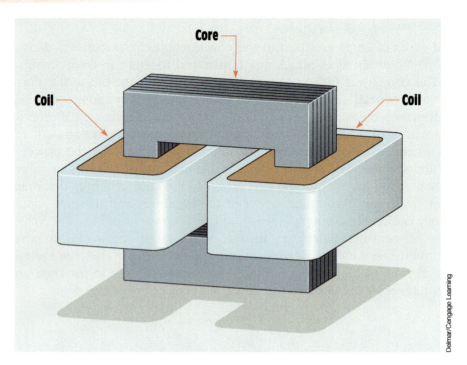

FIGURE 27–29 A core-type transformer.

The shell-type transformer is constructed in a similar manner to the core type, except that the shell type has a metal core piece through the middle of the window *(Figure 27–30)*. The primary and secondary windings are wound around the center core piece with the low-voltage winding being closest to the metal core. This arrangement permits the transformer to be surrounded by the core and provides excellent magnetic coupling. When the transformer is in operation, all the magnetic flux must pass through the center core piece. It then divides through the two outer core pieces.

The H-type core shown in *Figure 27–31* is similar to the shell-type core in that it has an iron core through its center around which the primary and secondary windings are wound. The H core, however, surrounds the windings on four sides instead of two. This extra metal helps reduce stray leakage flux and improves the efficiency of the transformer. The H-type core is often found on high-voltage distribution transformers.

The **tape-wound core** or **toroid core** *(Figure 27–32)* is constructed by tightly winding one long continuous silicon steel tape into a spiral. The tape may or may not be housed in a plastic container, depending on the application. This type of core does not require steel punchings laminated together. Because the core is one continuous length of metal, **flux leakage** is kept to a

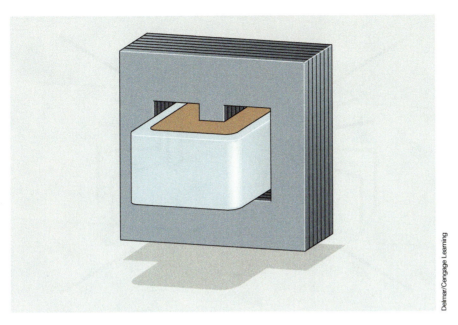

FIGURE 27–30 A shell-type transformer.

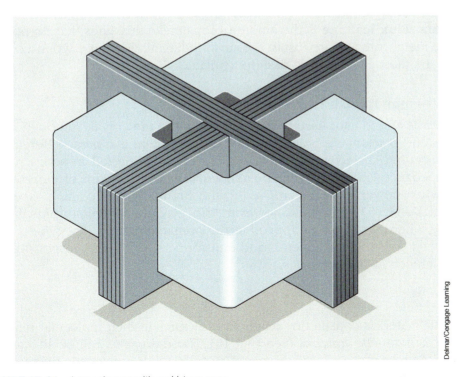

FIGURE 27–31 A transformer with an H-type core.

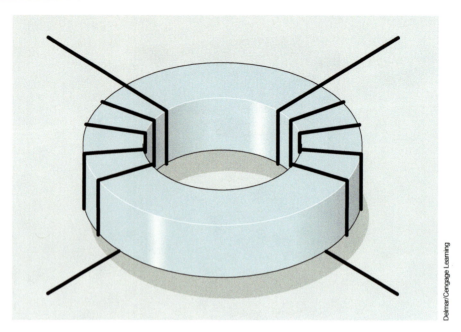

FIGURE 27–32 A toroid transformer.

minimum. **Flux leakage** is the amount of magnetic flux lines that do not follow the metal core and are lost to the surrounding air. The tape-wound core is one of the most efficient core designs available.

Transformer Inrush Current

A reactor is an inductor used to add inductance to the circuit. Although transformers and reactors are both inductive devices, there is a great difference in their operating characteristics. Reactors are often connected in series with a low-impedance load to prevent **inrush current** (the amount of current that flows when power is initially applied to the circuit) from becoming excessive *(Figure 27–33)*. Transformers, however, can produce extremely high inrush currents when power is first applied to the primary winding. The type of core used when constructing inductors and transformers is primarily responsible for this difference in characteristics.

Magnetic Domains

Magnetic materials contain tiny magnetic structures in their molecular material known as *magnetic domains* (See Unit 10). These domains can be affected by outside sources of magnetism. *Figure 27–34* illustrates a magnetic domain that has not been polarized by an outside magnetic source.

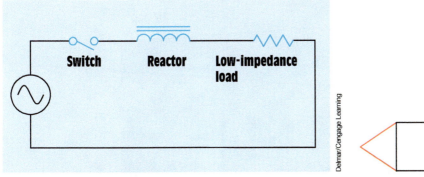

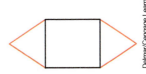

FIGURE 27–33 Reactors are used to help prevent inrush current from becoming excessive when power is first turned on.

FIGURE 27–34 Magnetic domain in neutral position.

Now assume that the north pole of a magnet is placed toward the top of the material that contains the magnetic domains *(Figure 27–35)*. Notice that the structure of the domain has changed to realign the molecules in the direction of the outside magnetic field. If the polarity of the magnetic pole is changed *(Figure 27–36)*, the molecular structure of the domain changes to realign itself with the new magnetic lines of flux. This external influence can be produced by an electromagnet as well as a permanent magnet.

In certain types of cores, the molecular structure of the domain snaps back to its neutral position when the magnetizing force is removed. This type of core is used in the construction of reactors or chokes *(Figure 27–37)*. A core of this type is constructed by separating sections of the steel laminations with

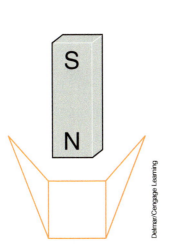

FIGURE 27–35 Domain influenced by a north magnetic field.

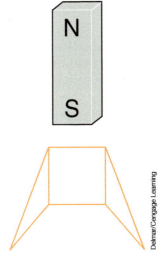

FIGURE 27–36 Domain influenced by a south magnetic field.

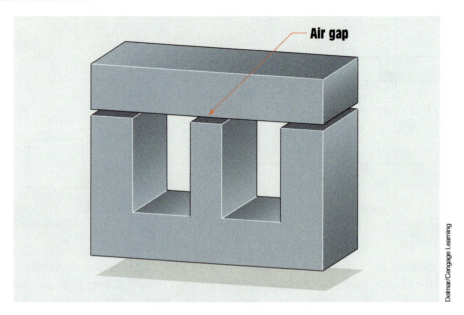

FIGURE 27–37 The core of an inductor contains an air gap.

an air gap. This air gap breaks the magnetic path through the core material and is responsible for the domains returning to their neutral position once the magnetizing force is removed.

The core construction of a transformer, however, does not contain an air gap. The steel laminations are connected together in such a manner as to produce a very low reluctance path for the magnetic lines of flux. In this type of core, the domains remain in their set position once the magnetizing force has been removed. This type of core "remembers" where it was last set. This was the principle of operation of the core memory of early computers. It is also the reason that transformers can have extremely high inrush currents when they are first connected to the powerline.

The amount of inrush current in the primary of a transformer is limited by three factors:

1. the amount of applied voltage,
2. the resistance of the wire in the primary winding, and
3. the flux change of the magnetic field in the core. The amount of flux change determines the amount of inductive reactance produced in the primary winding when power is applied.

Figure 28–38 illustrates a simple isolation-type transformer. The AC applied to the primary winding produces a magnetic field around the winding. As the current changes in magnitude and direction, the magnetic lines of flux change also. Because the lines of flux in the core are continually changing polarity,

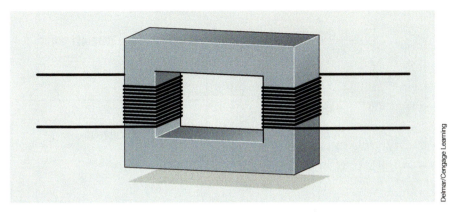

FIGURE 27-38 Isolation transformer.

the magnetic domains in the core material are changing also. As stated previously, the magnetic domains in the core of a transformer remember their last set position. For this reason, the point on the waveform at which current is disconnected from the primary winding can have a great bearing on the amount of inrush current when the transformer is reconnected to power. For example, assume the power supplying the primary winding is disconnected at the zero crossing point *(Figure 27-39)*. In this instance, the magnetic domains would be set at the neutral point. When power is restored to the primary winding, the core material can be magnetized by either magnetic polarity. This permits a change of flux, which is the dominant current-limiting factor. In this instance, the amount of inrush current would be relatively low.

If the power supplying current to the primary winding is interrupted at the peak point of the positive or negative half cycle, however, the domains in the core material will be set at that position. *Figure 27-40* illustrates this condition. It is assumed that the current was stopped as it reached its peak positive point. If

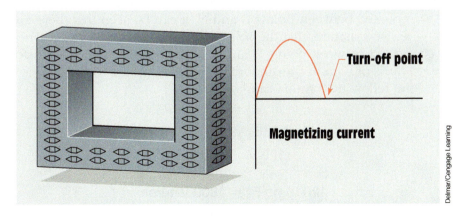

FIGURE 27-39 Magnetic domains are left in the neutral position.

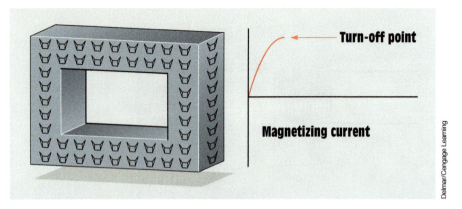

FIGURE 27–40 Domains are set at one end of magnetic polarity.

the power is reconnected to the primary winding during the positive half cycle, only a very small amount of flux change can take place. Because the core material is saturated in the positive direction, the primary winding of the transformer is essentially an air-core inductor, which greatly decreases the inductive characteristics of the winding. The inrush current in this situation would be limited by the resistance of the winding and a very small amount of inductive reactance.

This characteristic of transformers can be demonstrated with a clamp-on ammeter that has a "peak-hold" capability. If the ammeter is connected to one of the primary leads and power is switched on and off several times, the amount of inrush current varies over a wide range.

27–3 Autotransformers

Autotransformers are one-winding transformers. They use the same winding for both the primary and secondary. The primary winding in *Figure 27–41* is between points B and N and has a voltage of 120 volts applied to it. If the turns of wire are counted between points B and N, it can be seen that there are 120 turns of wire. Now assume that the selector switch is set to point D. The load is now connected between points D and N. The secondary of this transformer contains 40 turns of wire. If the amount of voltage applied to the load is to be calculated the following formula can be used:

$$\frac{E_P}{E_S} = \frac{N_P}{N_S}$$

$$\frac{120 \text{ V}}{E_S} = \frac{120 \text{ turns}}{40 \text{ turns}}$$

$$120 \text{ turns } E_S = 4800 \text{ V-turns}$$

$$E_S = 40 \text{ V}$$

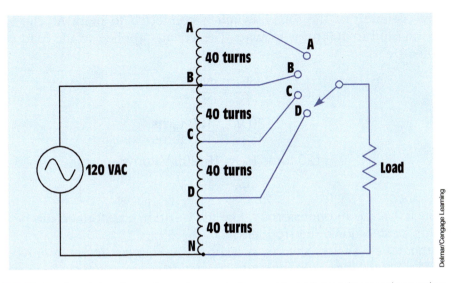

FIGURE 27–41 An autotransformer has only one winding used for both the primary and secondary.

Assume that the load connected to the secondary has an impedance of 10 ohms. The amount of current flow in the secondary circuit can be calculated using the formula

$$I = \frac{E}{Z}$$
$$I = \frac{40 \text{ V}}{10 \text{ }\Omega}$$
$$I = 4 \text{ A}$$

The primary current can be calculated by using the same formula that was used to calculate primary current for an isolation type of transformer:

$$\frac{E_P}{E_S} = \frac{I_S}{I_P}$$
$$\frac{120 \text{ V}}{40 \text{ V}} = \frac{4 \text{ A}}{I_P}$$
$$120 \text{ V } I_P = 160 \text{ VA}$$
$$I_P = 1.333 \text{ A}$$

The amount of power input and output for the autotransformer must be the same, just as they are in an isolation transformer:

Primary	Secondary
120 V × 1.333 A = 160 VA	40 V × 4 A = 160 VA

Now assume that the rotary switch is connected to point A. The load is now connected to 160 turns of wire. The voltage applied to the load can be calculated by

$$\frac{E_P}{E_S} = \frac{N_P}{N_S}$$

$$\frac{120\ V}{E_S} = \frac{120\ turns}{160\ turns}$$

$$120\ turns\ E_S = 19,200\ V\text{-turns}$$

$$E_S = 160\ V$$

Notice that the autotransformer, like the isolation transformer, can be either a step-up or step-down transformer.

If the rotary switch shown in *Figure 27–41* were to be removed and replaced with a sliding tap that made contact directly to the transformer winding, the turns ratio could be adjusted continuously. This type of transformer is commonly referred to as a Variac or Powerstat depending on the manufacturer. A cutaway view of a variable autotransformer is shown in *Figure 27–42*. The windings are wrapped around a tape-wound toroid core inside a plastic case. The tops of the windings have been milled flat to provide a commutator. A carbon brush makes contact with the windings.

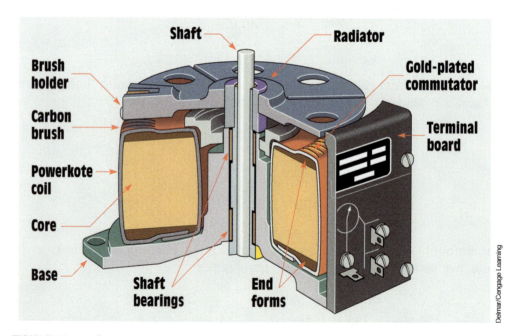

FIGURE 27–42   Cutaway view of a Powerstat.

FIGURE 27-43 Three-phase autotransformer.

Autotransformers are often used by power companies to provide a small increase or decrease to the line voltage. They help provide voltage regulation to large powerlines. A three-phase autotransformer is shown in *Figure 27–43*. This transformer is contained in a housing filled with transformer oil, which acts as a coolant and prevents moisture from forming in the windings.

The autotransformer does have one disadvantage. Because the load is connected to one side of the powerline, there is no line isolation between the incoming power and the load. This can cause problems with certain types of equipment and must be a consideration when designing a power system.

27-4 Transformer Polarities

To understand what is meant by transformer polarity, the voltage produced across a winding must be considered during some point in time. In a 60-herz AC circuit, the voltage changes polarity 60 times per second. When discussing transformer polarity, it is necessary to consider the relationship between the different windings at the same point in time. It is therefore assumed that this point in time is when the peak positive voltage is being produced across the winding.

Polarity Markings on Schematics

When a transformer is shown on a schematic diagram, it is common practice to indicate the polarity of the transformer windings by placing a dot beside one end of each winding as shown in *Figure 27–44*. These dots signify that the polarity is the same at that point in time for each winding. For example, assume the voltage applied to the primary winding is at its peak positive value at the terminal indicated by the dot. The voltage at the dotted lead of the secondary will be at its peak positive value at the same time.

This same type of polarity notation is used for transformers that have more than one primary or secondary winding. An example of a transformer with a multisecondary is shown in *Figure 27–45*.

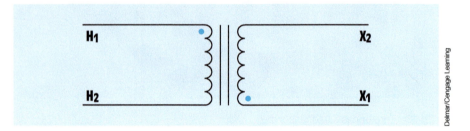

FIGURE 27–44 Transformer polarity dots.

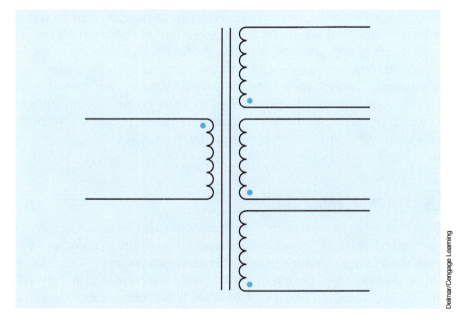

FIGURE 27–45 Polarity marks for multiple secondaries.

Additive and Subtractive Polarities

The polarity of transformer windings can be determined by connecting them as an autotransformer and testing for additive or subtractive polarity, often referred to as a boost or buck connection. This is done by connecting one lead of the secondary to one lead of the primary and measuring the voltage across both windings *(Figure 27–46)*. The transformer shown in the example has a primary voltage rating of 120 volts and a secondary voltage rating of 24 volts. This same circuit has been redrawn in *Figure 27–47* to show the connection more clearly. Notice that the secondary winding has been connected in series with the primary winding. The transformer now contains only one

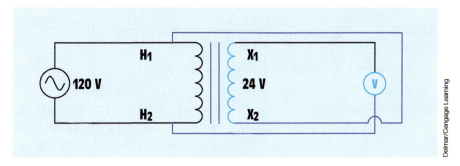

FIGURE 27–46 Connecting the secondary and primary windings forms an autotransformer.

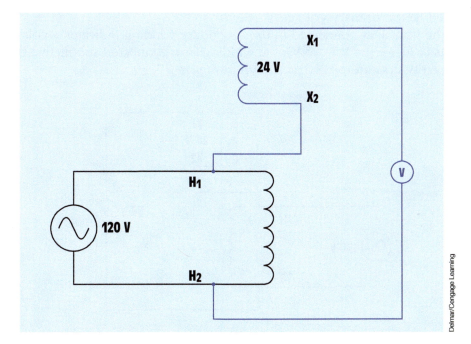

FIGURE 27–47 Redrawing the connection.

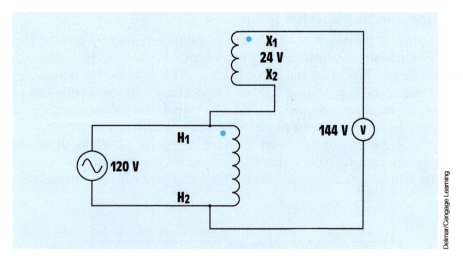

FIGURE 27–48 Placing polarity dots to indicate additive polarity.

winding and is therefore an autotransformer. When 120 volts are applied to the primary winding, the voltmeter connected across the secondary indicates either the *sum* of the two voltages or the *difference* between the two voltages. If this voltmeter indicates 144 volts (120 V + 24 V = 144 V), the windings are connected additive (boost) and polarity dots can be placed as shown in *Figure 27–48*. Notice in this connection that the secondary voltage is added to the primary voltage.

If the voltmeter connected to the secondary winding indicates a voltage of 96 volts (120 V − 24 V = 96 V), the windings are connected subtractive (buck) and polarity dots are placed as shown in *Figure 27–49*.

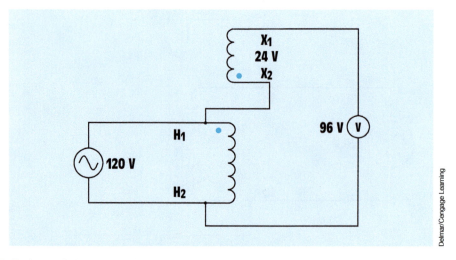

FIGURE 27–49 Polarity dots indicate subtractive polarity.

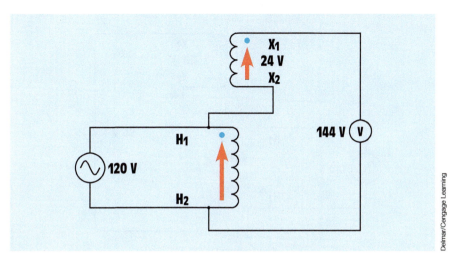

FIGURE 27–50 Arrows help indicate the placement of the polarity dots.

Using Arrows to Place Dots

To help in the understanding of additive and subtractive polarity, arrows can be used to indicate a direction of greater-than or less-than values. In *Figure 27–50*, arrows have been added to indicate the direction in which the dot is to be placed. In this example, the transformer is connected additive, or boost, and both arrows point in the same direction. Notice that the arrow points to the dot. In *Figure 27–51*, it is seen that values of the two arrows add to produce 144 volts.

In *Figure 27–52*, arrows have been added to a subtractive, or buck, connection. In this instance, the arrows point in opposite directions and the voltage of one tries to cancel the voltage of the other. The result is that the smaller value is eliminated and the larger value is reduced as shown in *Figure 27–53*.

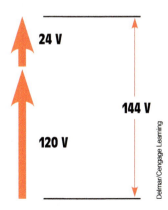

FIGURE 27–51 The values of the arrows add to indicate additive polarity (boost connection).

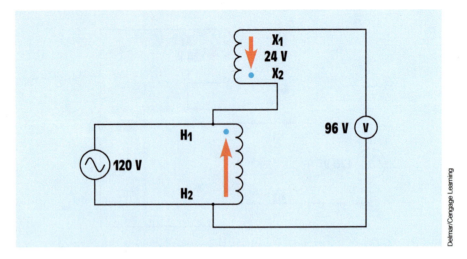

FIGURE 27–52 The arrows help indicate subtractive polarity.

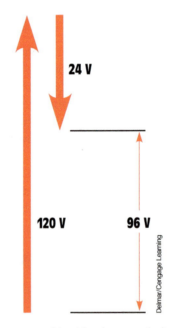

FIGURE 27–53 The values of the arrows subtract (buck connection).

27–5 Voltage and Current Relationships in a Transformer

When the primary of a transformer is connected to power but there is no load connected to the secondary, current is limited by the inductive reactance of the primary. At this time, the transformer is essentially an inductor and the excitation current is lagging the applied voltage by 90° *(Figure 27–54)*.

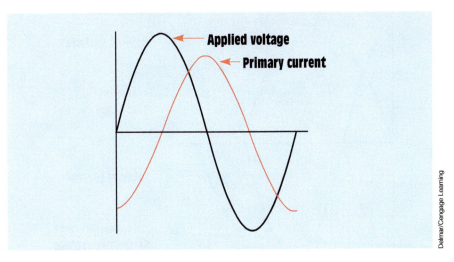

FIGURE 27–54 At no load, the primary current lags the voltage by 90°.

The primary current induces a voltage in the secondary. This induced voltage is proportional to the rate of change of current. The secondary voltage is maximum during the periods that the primary current is changing the most (0°, 180°, and 360°), and it will be zero when the primary current is not changing (90° and 270°). A plot of the primary current and secondary voltage shows that the secondary voltage lags the primary current by 90° *(Figure 27–55)*. Because the secondary voltage lags the primary current by 90° and the applied voltage leads the primary current by 90°, the secondary voltage is 180° out of phase with the applied voltage and in phase with the induced voltage in the primary.

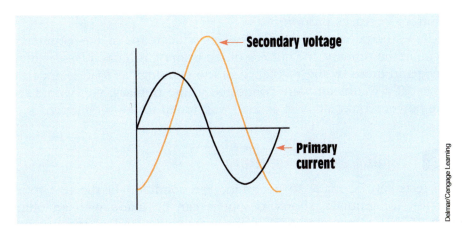

FIGURE 27–55 The secondary voltage lags the primary current by 90°.

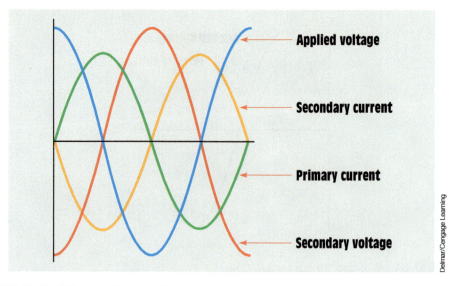

FIGURE 27-56 Voltage and current relationships of the primary and secondary windings.

Adding Load to the Secondary

When a load is connected to the secondary, current begins to flow. Because the transformer is an inductive device, the secondary current lags the secondary voltage by 90°. Because the secondary voltage lags the primary current by 90°, the secondary current is 180° out of phase with the primary current *(Figure 27-56)*.

The current of the secondary induces a countervoltage in the secondary windings that is in opposition to the countervoltage induced in the primary. The countervoltage of the secondary weakens the countervoltage of the primary and permits more primary current to flow. As secondary current increases, primary current increases proportionally.

Because the secondary current causes a decrease in the countervoltage produced in the primary, the current of the primary is limited less by inductive reactance and more by the resistance of the windings as load is added to the secondary. If a wattmeter were connected to the primary, you would see that the true power would increase as load was added to the secondary.

27-6 Testing the Transformer

Several tests can be made to determine the condition of the transformer. A simple test for grounds, shorts, or opens can be made with an ohmmeter *(Figure 27-57)*. Ohmmeter A is connected to one lead of the primary and one lead of the secondary. This test checks for shorted windings between the

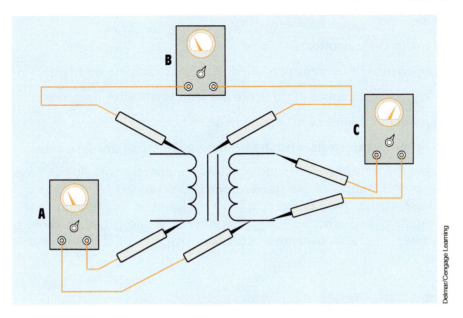

FIGURE 27–57 Testing a transformer with an ohmmeter.

primary and secondary. The ohmmeter should indicate infinity. If there is more than one primary or secondary winding, all isolated windings should be tested for shorts. Ohmmeter B illustrates testing the windings for grounds. One lead of the ohmmeter is connected to the case of the transformer, and the other is connected to the winding. All windings should be tested for grounds, and the ohmmeter should indicate infinity for each winding. Ohmmeter C illustrates testing the windings for continuity. The wire resistance of the winding should be indicated by the ohmmeter.

If the transformer appears to be in good condition after the ohmmeter test, it should then be tested for shorts and grounds with a megohmmeter. A MEGGER® will reveal problems of insulation breakdown that an ohmmeter will not. Large oil-filled transformers should have the condition of the dielectric oil tested at periodic intervals. This test involves taking a sample of the oil and performing certain tests for dielectric strength and contamination.

27–7 Transformer Nameplates

Most transformers contain a nameplate that lists information concerning the transformer. *NEC® 450.11* requires the following information:

1. Name of manufacturer
2. Rated kilovolt-ampere

3. Frequency
4. Primary and secondary voltage
5. Impedance of transformers rated 25 kilovolt-ampere and larger
6. Required clearances of transformers with ventilating openings
7. Amount and kind of insulating liquid where used
8. The temperature class for the insulating system of dry-type transformers

Notice that the transformer is rated in kilovolt-amperes, not kilowatts, because the true power output of the transformer is determined by the power factor of the load. Other information that may be listed is temperature rise in °C, model number, and whether the transformer is single phase or three phase. Many nameplates also contain a schematic diagram of the windings to aid in connection.

27–8 Determining Maximum Current

The nameplate does not list the current rating of the windings. Because power input must equal power output, the current rating for a winding can be determined by dividing the kilovolt-ampere rating by the winding voltage. For example, assume a transformer has a kilovolt-ampere rating of 0.5 kilovolt-ampere, a primary voltage of 480 volts, and a secondary voltage of 120 volts. To determine the maximum current that can be supplied by the secondary, divide the kVA rating by the secondary voltage:

$$I_S = \frac{kVA}{E_S}$$

$$I_S = \frac{500 \text{ VA}}{120 \text{ V}}$$

$$I_S = 4.167 \text{ A}$$

The primary current can be calculated in the same way:

$$I_P = \frac{kVA}{E_P}$$

$$I_P = \frac{500 \text{ VA}}{480 \text{ V}}$$

$$I_P = 1.042 \text{ A}$$

Transformers with multiple secondary windings will generally have the current rating listed with the voltage rating.

27-9 Transformer Impedance

Transformer impedance is determined by the physical construction of the transformer. Factors such as the amount and type of core material, wire size used to construct the windings, the number of turns, and the degree of magnetic coupling between the windings greatly affect the transformer's impedance. Impedance is expressed as a percent (%Z or %IZ) and is measured by connecting a short circuit across the low-voltage winding of the transformer and then connecting a variable voltage source to the high-voltage winding *(Figure 27–58)*. The variable voltage is then increased until rated current flows in the low-voltage winding. The transformer impedance is determined by calculating the percentage of variable voltage as compared to the rated voltage of the high-voltage winding.

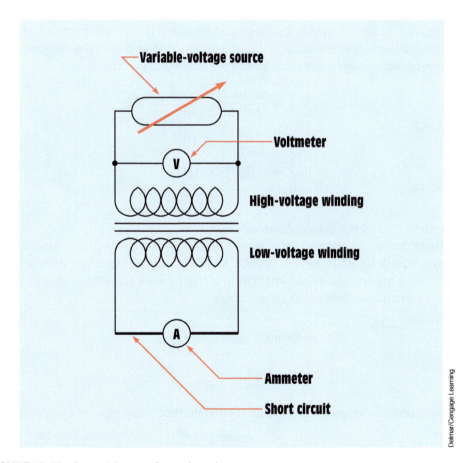

FIGURE 27–58 Determining transformer impedance.

■ EXAMPLE 27-1

Assume that the transformer shown in *Figure 27–58* is a 2400/480-V, 15-kVA transformer. To determine the impedance of the transformer, first calculate the full-load current rating of the secondary winding:

$$I = \frac{VA}{E}$$

$$I = \frac{15,000 \text{ VA}}{480 \text{ V}}$$

$$I = 31.25 \text{ A}$$

Next, increase the source voltage connected to the high-voltage winding until a current of 31.25 A flows in the low-voltage winding. For the purpose of this example, assume that the voltage value is 138 V. Finally, determine the percentage of applied voltage as compared to the rated voltage:

$$\%Z = \frac{\text{source voltage}}{\text{rated voltage}} \times 100$$

$$\%Z = \frac{138 \text{ V}}{2400 \text{ V}} \times 100$$

$$\%Z = 0.0575 \times 100$$

$$\%Z = 5.75$$

The impedance of this transformer is 5.75%.

Transformer impedance is a major factor in determining the amount of voltage drop a transformer will exhibit between no load and full load and in determining the amount of current flow in a short-circuit condition. Short-circuit current can be calculated using the formula

$$\text{(Single phase) } I_{SC} = \frac{VA}{E \times \%Z}$$

$$\text{(Three phase) } I_{SC} = \frac{VA}{E \times \sqrt{3} \times \%Z}$$

because one of the formulas for determining current in a single-phase circuit is

$$I = \frac{VA}{E}$$

and one of the formulas for determining current in a three-phase circuit is

$$I = \frac{VA}{E \times \sqrt{3}}$$

The preceding formulas for determining short-circuit current can be modified to show that the short-circuit current can be calculated by dividing the rated secondary current by the %Z:

$$I_{SC} = \frac{I_{Rated}}{\%Z}$$

EXAMPLE 27-2

A single-phase transformer is rated at 50 kVA and has a secondary voltage of 240 V. The nameplate reveals that the transformer has an internal impedance (%Z) of 2.5%. What is the short-circuit current for this transformer?

$$I_{secondary} = \frac{50{,}000 \text{ VA}}{240 \text{ V}}$$

$$I_{secondary} = 208.3 \text{ A}$$

$$I_{short\ circuit} = \frac{208.3 \text{ A}}{\%Z}$$

$$I_{short\ circuit} = \frac{208.3 \text{ A}}{0.025}$$

$$I_{short\ circuit} = 8{,}333.3 \text{ A}$$

It is sometimes necessary to calculate the amount of short-circuit current when determining the correct fuse rating for a circuit. The fuse must have a high enough "interrupt" rating to clear the fault in the event of a short circuit.

Constant-Current Transformers

A very special type of isolation transformer is the **constant-current transformer,** often referred to as a **current regulator.** Constant-current transformers are designed to deliver a constant output current, generally 6.6 amperes, under varying load conditions. They are most often employed to provide power to series-connected street lamps. Street lamps are often connected in series instead of parallel because of the savings in wire. Series-connected lamps require a single conductor to be connected from lamp to lamp instead of two conductors, *Figure 27–59*.

When lamps are connected in series, some device must be used to continue the circuit in the event that one or more lamps should fail. Some lights use a reactor coil connected in parallel with the lamp, *Figure 27–60*. Another method uses a film cut-out device, *Figure 27–61*, consisting of two pieces of metal separated by an insulator designed to puncture at a predetermined voltage. As long as the lamp is in operation, the voltage drop across the cut-out device is not sufficient to cause the film to puncture. If the lamp should burn out, producing an open circuit, the entire circuit voltage will appear across the cut-out device and cause it to short circuit.

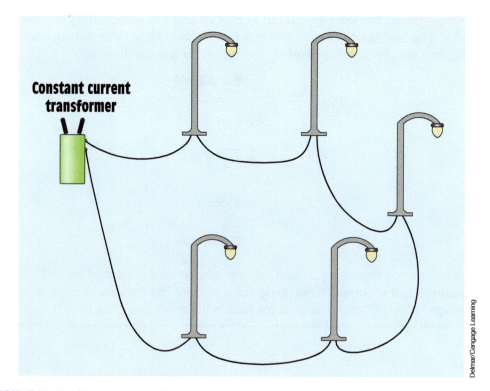

FIGURE 27–59 Street lamps are often connected in series.

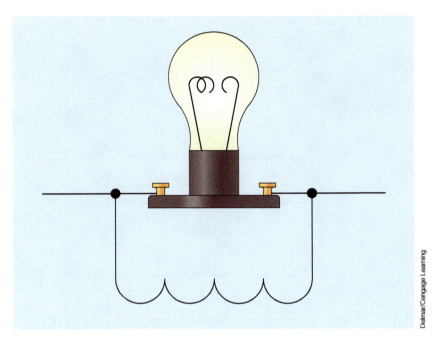

FIGURE 27–60 An inductor maintains the circuit if the lamp should fail.

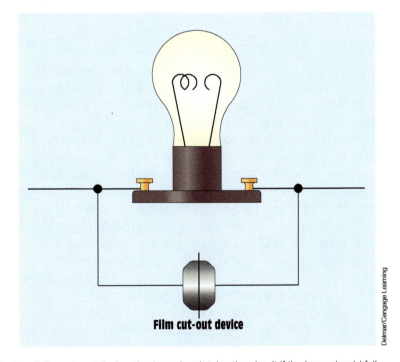

FIGURE 27–61 A film cut-out device shorts and maintains the circuit if the lamp should fail.

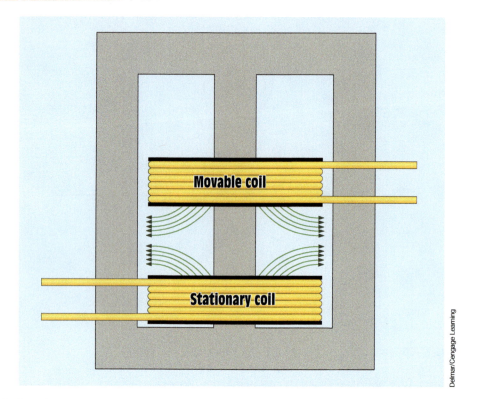

FIGURE 27-62 Magnetic flux of the two windings repel each other.

Constant-current transformers contain primary and secondary winding that are movable with respect to each other. Either winding can be made movable. Both windings are wound on the same core material, *Figure 27-62*.

The constant-current regulator operates by producing a magnetic field in the movable winding that is the same polarity as the magnetic field produced in the stationary winding. Because the two magnetic fields have the same polarity, they oppose each other. If the load current increases, the magnetic field strength of the two windings increases, causing the two coils to move further apart. Moving coils farther apart increases the amount of magnetic flux leakage, resulting in a reduction in output voltage. If the load current decreases, the magnetic field strength of the two windings decreases, causing the movable coil to move closer to the stationary coil, producing an increase in output voltage.

Many constant-current transformers employ a counterweight and dashpot mechanism to help reduce sudden changes in the spacing between the two coils, *Figure 27-63*. The counterweight helps balance the weight of the movable coil, and the dashpot device helps reduce the "hunting" action between the two coils.

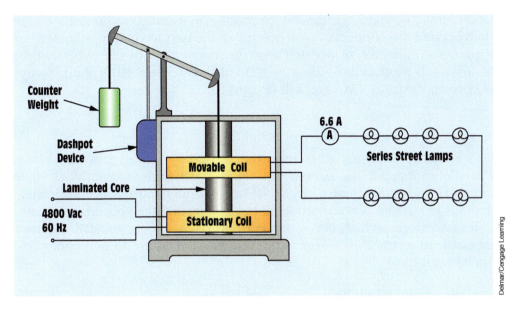

FIGURE 27–63 A counterweight and dashpot device help reduce sudden changes in the output current.

Series Connection of Transformer Secondaries

As a general rule, connecting the secondary windings of transformers in series does not present a problem. Because the current is the same in a series circuit, the problem of high circulating current does not exist in a series connection. The secondary windings can be connected in series to produce a higher output voltage. Assume that two transformers have a secondary voltage of 120 volts, *Figure 27–64*.

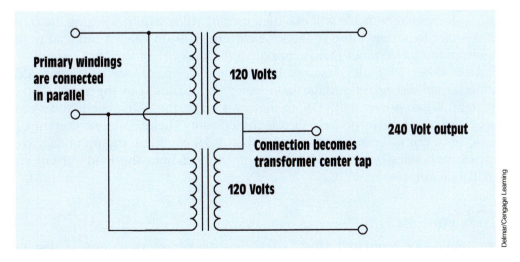

FIGURE 27–64 Transformer secondary windings connected in series.

The primary windings can be connected in parallel without a problem. When making this connection, the polarity of the two secondary windings must be connected additive of boost. The series connection of the two secondary windings will produce an output of 240 volts center tapped. If the polarity is not correct, the output voltage will be zero (0).

Parallel Transformer Connections

It is sometimes necessary to connect the secondary windings of transformers in parallel to increase the current capacity, but generally it is not done unless there is no other alternative, *Figure 27–65*. Connecting transformer primary windings in parallel is not a problem, but connecting the secondary windings in parallel can cause high circulating currents or extremely unbalance currents that can lead to transformer failure. Transformers that are to be connected in parallel must have

- the same kVA rating.
- the same turns ratio,
- the same impedance, and
- the same secondary voltage;
- and the secondary windings must have the same polarity.

If any of these factors are different, it can cause failure of one or both transformers. Assume for example, that two transformers have the same kVA rating, same secondary voltage, and same turns ratio, but the impedance is not the same. When load is added to the parallel connection, the transformer with the higher internal impedance will exhibit a greater voltage drop, causing the other transformer to supply more of the load current. This unbalance can lead to the failure of the transformer that is supplying the majority of the load current.

In another example, assume that two transformers have the same kVA rating, same secondary voltage, and same impedance, but the turns ratio is different. When power is applied to the connection, the difference in turns ratio can cause a very high no-load circulating current. This circulating current can cause burn-out of both transformers. When load is added, the secondary current of each transformer will be a combination of both the load current and circulating current.

Connecting Parallel Transformers

Care should be exercised when paralleling transformers to ensure that the polarity is correct. If the polarity is incorrect, it produces a short circuit. In this

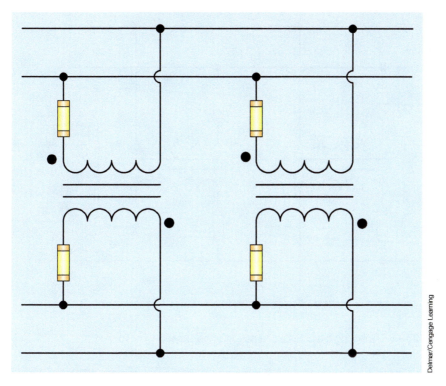

FIGURE 27–65 Transformers are sometimes connected in parallel to increase the current capacity of the circuit.

example, assume that one transformer is already connected to the line. Also assumed that the primary voltage is 4160 volts and the secondary voltage is 120 volts. Connect one of the secondary leads to the line and then energize the primary winding, *Figure 27–66*.

Connect a voltmeter between the secondary lead that has not been connected and the line of the intended connection. If the polarity of the two transformers is correct, the voltmeter should indicate zero (0) volts. If the polarity is not correct, the voltmeter will indicate double the amount of secondary voltage. In this example, the voltmeter would indicate 240 volts if the polarity were not correct.

Precautions When Servicing Parallel Transformers

If it should become necessary to remove one of the transformers for service, the secondary should be disconnected first. This can generally be accomplished by removing the secondary fuse, *Figure 27–67*.

In this example, the primary is connected to 4160 volts. If the secondary winding is not disconnected first, the transformer becomes a step-up

810 SECTION XIII Transformers

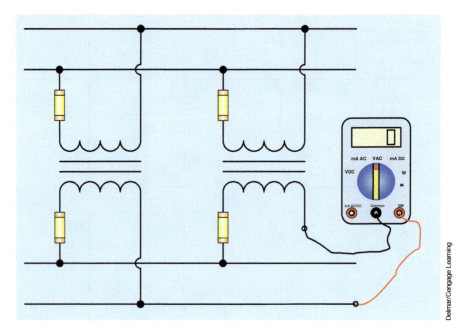

FIGURE 27–66 Checking the polarity of a parallel transformer connection.

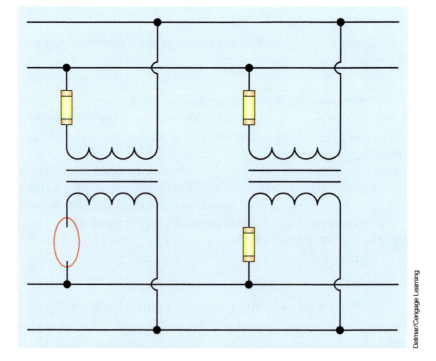

FIGURE 27–67 The secondary winding should be disconnected before disconnecting the primary winding.

transformer. The primary winding will still have a voltage of 4160 even if it is disconnected from the power line. Anytime parallel transformers are employed, a sign reading **CAUTION FEEDBACK VOLTAGE** should be located at each primary fuse.

Summary

- All values of voltage, current, and impedance in a transformer are proportional to the turns ratio.
- Transformers can change values of voltage, current, and impedance but cannot change the frequency.
- The primary winding of a transformer is connected to the powerline.
- The secondary winding is connected to the load.
- A transformer that has a lower secondary voltage than primary voltage is a step-down transformer.
- A transformer that has a higher secondary voltage than primary voltage is a step-up transformer.
- An isolation transformer has its primary and secondary windings electrically and mechanically separated from each other.
- When a coil induces a voltage into itself, it is known as self-induction.
- When a coil induces a voltage into another coil, it is known as mutual induction.
- Transformers can have very high inrush current when first connected to the powerline because of the magnetic domains in the core material.
- Inductors provide an air gap in their core material that causes the magnetic domains to reset to a neutral position.
- Autotransformers have only one winding, which is used as both the primary and secondary.
- Autotransformers have a disadvantage in that they have no line isolation between the primary and secondary winding.
- Isolation transformers help filter voltage and current spikes between the primary and secondary side.
- Polarity dots are often added to schematic diagrams to indicate transformer polarity.

- Transformers can be connected as additive or subtractive polarity.
- Constant-current transformers are also known as current regulators.
- Constant-current transformers are generally used to provide power to series-connected loads.
- As a general rule, transformer secondary windings should not be connected in parallel.

Review Questions

1. What is a transformer?
2. What are common efficiencies for transformers?
3. What is an isolation transformer?
4. All values of a transformer are proportional to its _____ _____.
5. What is an autotransformer?
6. What is a disadvantage of an autotransformer?
7. Explain the difference between a step-up and a step-down transformer.
8. A transformer has a primary voltage of 240 V and a secondary voltage of 48 V. What is the turns ratio of this transformer?
9. A transformer has an output of 750 VA. The primary voltage is 120 V. What is the primary current?
10. A transformer has a turns ratio of 1:6. The primary current is 18 A. What is the secondary current?
11. What do the dots shown beside the terminal leads of a transformer represent on a schematic?
12. A transformer has a primary voltage rating of 240 V and a secondary voltage rating of 80 V. If the windings were connected subtractive, what voltage would appear across the entire connection?
13. If the windings of the transformer in Question 12 were to be connected additive, what voltage would appear across the entire winding?
14. The primary leads of a transformer are labeled 1 and 2. The secondary leads are labeled 3 and 4. If polarity dots are placed beside leads 1 and 4, which secondary lead would be connected to terminal 2 to make the connection additive?

Practical Applications

You are working in an industrial plant. You must install a single-phase transformer. The transformer has the following information on the nameplate:

Primary voltage—13,800 V

Secondary voltage—240 V

Impedance—5%

150 kVA

The secondary fuse has a blow rating of 800 A and an interrupt rating of 10,000 A. Is this interrupt rating sufficient for this installation? ■

Practical Applications

You have been given a transformer to install on a 277-V line. The transformer nameplate is shown in *Figure 27–68*. The transformer must supply a 120-V, 20-A circuit. The transformer capacity should be not less than 115% of the rated load. Does the transformer you have been given have enough kVA capacity to supply the load? To which transformer terminals would you connect the incoming power? To which transformer terminals would you connect the load? ■

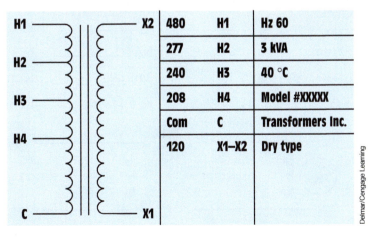

FIGURE 27–68 Transformer nameplate.

Practice Problems

Refer to *Figure 27–69* to answer the following questions. Find all the missing values.

1.
- E_P 120 V
- E_S 24 V
- I_P _____
- I_S _____
- N_P 300 turns
- N_S _____
- Ratio _____
- $Z = 3\ \Omega$

2.
- E_P 240 V
- E_S 320 V
- I_P _____
- I_S _____
- N_P _____
- N_S 280
- Ratio _____
- $Z = 500\ \Omega$

3.
- E_P _____
- E_S 160 V
- I_P _____
- I_S _____
- N_P _____
- N_S 80 turns
- Ratio 1:2.5
- $Z = 12\ \Omega$

4.
- E_P 48
- E_S 240 V
- I_P _____
- I_S _____
- N_P 220 turns
- N_S _____
- Ratio _____
- $Z = 360\ \Omega$

5.
- E_P _____
- E_S _____
- I_P 16.5 A
- I_S 3.25 A
- N_P _____
- N_S 450 turns
- Ratio _____
- $Z = 56\ \Omega$

6.
- E_P 480 V
- E_S _____
- I_P _____
- I_S _____
- N_P 275 turns
- N_S 525 turns
- Ratio _____
- $Z = 1.2\ k\Omega$

Refer to *Figure 27–70* to answer the following questions. Find all the missing values.

7.
- E_P 208 V
- E_{S1} 320 V
- E_{S2} 120 V
- E_{S3} 24 V
- I_P _____
- I_{S1} _____
- I_{S2} _____
- I_{S3} _____
- NP 800 turns
- N_{S1} _____
- N_{S2} _____
- N_{S3} _____
- Ratio 1:
- Ratio 2:
- Ratio 3:
- R_1 12 kΩ
- R_2 6 Ω
- R_3 8 Ω

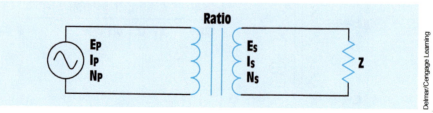

FIGURE 27–69 Isolation transformer practice problems.

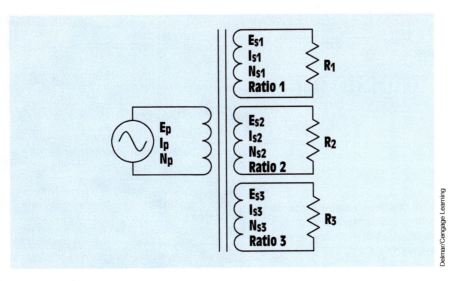

FIGURE 27–70 Single-phase transformer with multiple secondaries.

8.

E_P 277 V	E_{S1} 480 V	E_{S2} 208 V	E_{S3} 120 V
I_P _____	I_{S1} _____	I_{S2} _____	I_{S3} _____
N_P 350 turns	N_{S1} _____	N_{S2} _____	N_{S3} _____
	Ratio 1:	Ratio 2:	Ratio 3:
	R_1 200 Ω	R_2 60 Ω	R_3 24 Ω

Unit 28
Three-Phase Transformers

Why You Need to Know

Three-phase transformers are used throughout industry. Almost all power generated in the United States and Canada is three phase. Transformers step up voltage for transmission and step it down again for use inside a plant or commercial building. This unit

- presents the difference between a true three-phase transformer and a three-phase transformer bank.
- determines different voltage and current values in a three-phase transformer.
- defines phase values in calculating the values of a transformer.
- explains how harmonics are identified and overcome.
- describes different three-phase transformer connections such as delta–wye, wye–delta, open-delta, T connected, and Scott connected.
- discusses installation and testing.

OUTLINE

28–1	Three-Phase Transformers
28–2	Closing a Delta
28–3	Three-Phase Transformer Calculations
28–4	Open-Delta Connection
28–5	Single-Phase Loads
28–6	Closed Delta with Center Tap
28–7	Closed Delta without Center Tap
28–8	Delta–Wye Connection with Neutral
28–9	T-Connected Transformers
28–10	Scott Connection
28–11	Zig-Zag Connection
28–12	Harmonics

KEY TERMS

Closing a delta
Delta–wye
Dielectric oil
High leg
One-line diagram
Open-delta
Orange wire
Single-phase loads
Tagging
Three-phase bank
Wye–delta

Objectives

After studying this unit, you should be able to

- discuss the operation of three-phase transformers.
- connect three single-phase transformers to form a three-phase bank.
- calculate voltage and current values for a three-phase transformer connection.
- connect two single-phase transformers to form a three-phase open-delta connection.
- discuss the characteristics of an open-delta connection.
- discuss different types of three-phase transformer connections and how they are used to supply single-phase loads.
- calculate values of voltage and current for a three-phase transformer used to supply both three-phase and single-phase loads.
- describe what a harmonic is.
- discuss the problems concerning harmonics.
- identify the characteristics of different harmonics.
- perform a test to determine if harmonic problems exist.
- discuss methods of dealing with harmonic problems.

Preview

Three-phase transformers are used throughout industry to change values of three-phase voltage and current. Because three-phase power is the most common way in which power is produced, transmitted, and used, an understanding of how three-phase transformer connections are made is essential. This unit discusses different types of three-phase transformer connections and presents examples of how values of voltage and current for these connections are calculated. ■

28–1 Three-Phase Transformers

A three-phase transformer is constructed by winding three single-phase transformers on a single core *(Figure 28–1)*. A photograph of a three-phase transformer is shown in *Figure 28–2*. The transformer is shown before it is mounted in an enclosure, which will be filled with a **dielectric oil.** The dielectric oil performs several functions. Because it is a dielectric, it provides electric insulation between the windings and the case. It is also used to help provide cooling and to prevent the formation of moisture, which can deteriorate the winding insulation.

SECTION XIII Transformers

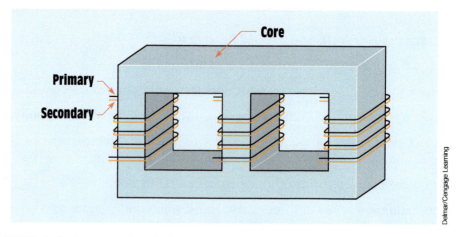

FIGURE 28–1 Basic construction of a three-phase transformer.

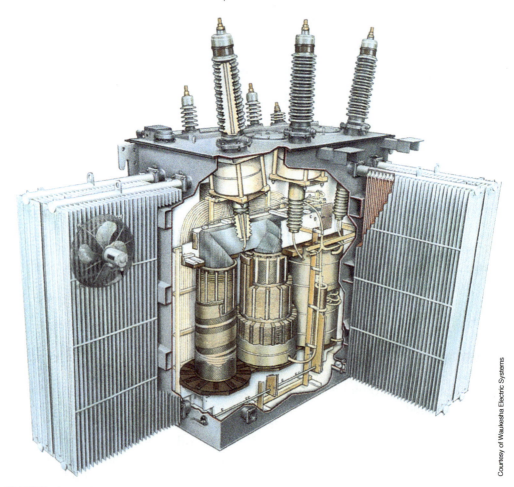

FIGURE 28–2 Three-phase transformer.

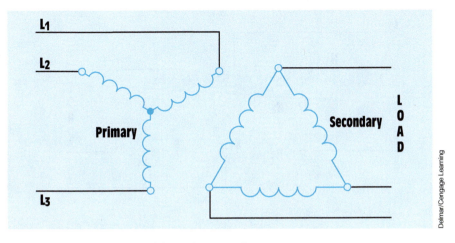

FIGURE 28–3 Wye–delta connected three-phase transformer.

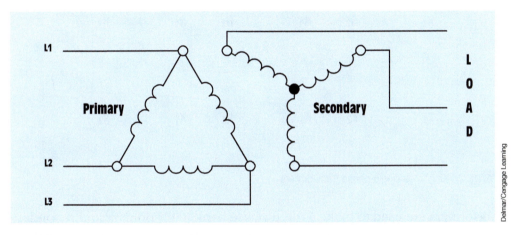

FIGURE 28–4 Delta–wye connected three-phase transformer.

Three-Phase Transformer Connections

Three-phase transformers are connected in delta or wye configurations. A **wye–delta** transformer, for example, has its primary winding connected in a wye and its secondary winding connected in a delta *(Figure 28–3)*. A **delta–wye** transformer would have its primary winding connected in delta and its secondary connected in wye *(Figure 28–4)*.

Connecting Single-Phase Transformers into a Three-Phase Bank

If three-phase transformation is needed and a three-phase transformer of the proper size and turns ratio is not available, three single-phase transformers can be connected to form a **three-phase bank.** When three single-phase

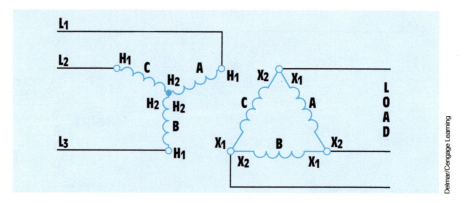

FIGURE 28–5 Identifying the windings.

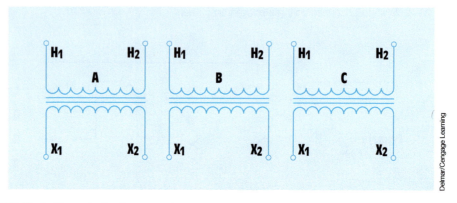

FIGURE 28–6 Three single-phase transformers.

transformers are used to make a three-phase bank, their primary and secondary windings are connected in a wye or delta connection. The three transformer windings in *Figure 28–5* are labeled A, B, and C. One end of each primary lead is labeled H_1, and the other end is labeled H_2. One end of each secondary lead is labeled X_1, and the other end is labeled X_2.

Figure 28–6 shows three single-phase transformers labeled A, B, and C. The primary leads of each transformer are labeled H_1 and H_2, and the secondary leads are labeled X_1 and X_2. The schematic diagram of *Figure 28–5* is used to connect the three single-phase transformers into a three-phase wye–delta connection, as shown in *Figure 28–7*.

The primary winding is first tied into a wye connection. The schematic in *Figure 28–5* shows that the H_2 leads of all three primary windings are connected together and the H_1 lead of each winding is open for connection to the incoming powerline. Notice in *Figure 28–7* that the H_2 leads of the primary windings are connected together and the H_1 lead of each winding has been connected to the incoming powerline.

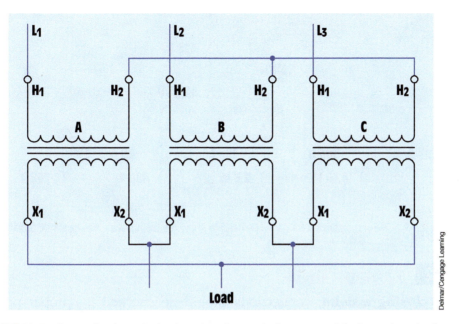

FIGURE 28–7 Connecting three single-phase transformers to form a wye–delta three-phase bank.

Figure 28–5 shows that the X_1 lead of Transformer A is connected to the X_2 lead of Transformer C. Notice that this same connection has been made in *Figure 28–7*. The X_1 lead of Transformer B is connected to the X_2 lead of Transformer A, and the X_1 lead of Transformer C is connected to the X_2 lead of Transformer B. The load is connected to the points of the delta connection.

Although *Figure 28–5* illustrates the proper schematic symbology for a three-phase transformer connection, some electrical schematics and wiring diagrams do not illustrate three-phase transformer connections in this manner. One type of diagram, called the **one-line diagram,** would illustrate a delta–wye connection, as shown in *Figure 28–8*. These diagrams are generally used to show the main power distribution system of a large industrial plant. The one-line diagram in *Figure 28–9* shows the main power to the plant and the transformation of voltages to different subfeeders. Notice that each transformer shows whether the primary and secondary are connected as a wye or delta and the secondary voltage of the subfeeder.

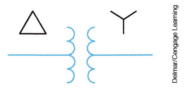

FIGURE 28–8 One-line diagram symbol used to represent a delta–wye three-phase transformer connection.

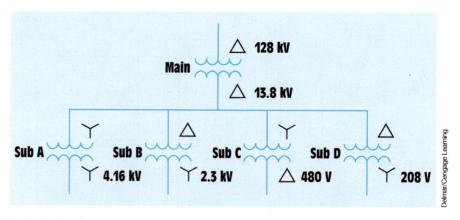

FIGURE 28–9 One-line diagrams are generally used to show the main power distribution of a plant.

28–2 Closing a Delta

When **closing a delta,** connections should be checked for proper polarity before making the final connection and applying power. If the phase winding of one transformer is reversed, an extremely high current will flow when power is applied. Proper phasing can be checked with a voltmeter, as shown in *Figure 28–10.* If power is applied to the transformer bank before the delta connection is closed, the voltmeter should indicate 0 volt. If one phase winding has been reversed, however, the voltmeter will indicate double the amount of voltage. For example, assume that the output voltage of a delta secondary is 240 volts. If the voltage is checked before the delta is closed, the voltmeter should indicate a voltage of 0 V if all windings have been phased properly. If one winding has been reversed, however, the voltmeter will indicate a voltage of 480 volts

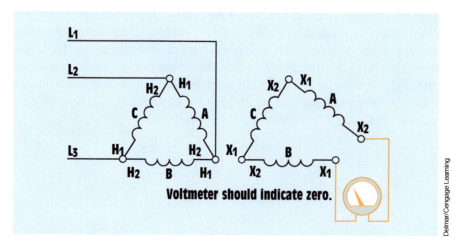

FIGURE 28–10 Testing for proper transformer polarity before closing the delta.

(240 V + 240 V). This test will confirm whether a phase winding has been reversed, but it will not indicate whether the reversed winding is located in the primary or secondary. If either primary or secondary windings have been reversed, the voltmeter will indicate double the output voltage.

Note, however, that a voltmeter is a high-impedance device. It is not unusual for a voltmeter to indicate some amount of voltage before the delta is closed, especially if the primary has been connected as a wye and the secondary as a delta. When this is the case, however, the voltmeter will generally indicate close to the normal output voltage if the connection is correct and double the output voltage if the connection is incorrect.

28–3 Three-Phase Transformer Calculations

To calculate the values of voltage and current for three-phase transformers, the formulas used for making transformer calculations and three-phase calculations must be followed. Another very important rule is that ***only phase values of voltage and current can be used when calculating transformer values.*** Refer to Transformer A in *Figure 28–6*. All transformation of voltage and current takes place between the primary and secondary windings. Because these windings form the phase values of the three-phase connection, only phase and not line values can be used when calculating transformed voltages and currents.

■ EXAMPLE 28–1

A three-phase transformer connection is shown in *Figure 28–11*. Three single-phase transformers have been connected to form a wye–delta bank. The primary is connected to a three-phase line of 13,800 V, and the secondary voltage is 480 V. A three-phase resistive load with an impedance of 2.77 Ω per phase is connected to the secondary of the transformer. Calculate the following values for this circuit:

$E_{P(PRIMARY)}$ — phase voltage of the primary

$E_{P(SECONDARY)}$ — phase voltage of the secondary

Ratio — turns ratio of the transformer

$E_{P(LOAD)}$ — phase voltage of the load bank

$I_{P(LOAD)}$ — phase current of the load bank

$I_{L(SECONDARY)}$ — secondary line current

$I_{P(SECONDARY)}$ — phase current of the secondary

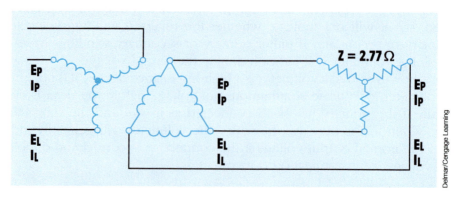

FIGURE 28–11 Example Circuit 1 three-phase transformer calculation.

$I_{P(PRIMARY)}$ — phase current of the primary

$I_{L(PRIMARY)}$ — line current of the primary

Solution

The primary windings of the three single-phase transformers have been connected to form a wye connection. In a wye connection, the phase voltage is less than the line voltage by a factor of 1.732 (the square root of 3). Therefore, the phase value of the primary voltage can be calculated using the formula

$$E_{P(PRIMARY)} = \frac{E_L}{1.732}$$

$$E_{P(PRIMARY)} = \frac{13{,}800 \text{ V}}{1.732}$$

$$E_{P(PRIMARY)} = 7967.667 \text{ V}$$

The secondary windings are connected as a delta. In a delta connection, the phase voltage and line voltage are the same:

$$E_{P(SECONDARY)} = E_{L(SECONDARY)}$$

$$E_{P(SECONDARY)} = 480 \text{ V}$$

The turns ratio can be calculated by comparing the phase voltage of the primary with the phase voltage of the secondary:

$$\text{Ratio} = \frac{\text{primary voltage}}{\text{secondary voltage}}$$

$$\text{Ratio} = \frac{7967.667 \text{ V}}{480 \text{ V}}$$

$$\text{Ratio} = 16.6:1$$

The load bank is connected in a wye connection. The voltage across the phase of the load bank will be less than the line voltage by a factor of 1.732:

$$E_{P(LOAD)} = \frac{E_{L(LOAD)}}{1.732}$$

$$E_{P(LOAD)} = \frac{480 \text{ V}}{1.732}$$

$$E_{P(LOAD)} = 277.136 \text{ V}$$

Now that the voltage across each of the load resistors is known, the current flow through the phase of the load can be calculated using Ohm's law:

$$I_{P(LOAD)} = \frac{E}{R}$$

$$I_{P(LOAD)} = \frac{277.136 \text{ V}}{2.77}$$

$$I_{P(LOAD)} = 100.049 \text{ A}$$

Because the load is connected as a wye connection, the line current is the same as the phase current:

$$I_{L(SECONDARY)} = 100.049 \text{ A}$$

The secondary of the transformer bank is connected as a delta. The phase current of the delta is less than the line current by a factor of 1.732:

$$I_{P(SECONDARY)} = \frac{I_L}{1.732}$$

$$I_{P(SECONDARY)} = \frac{100.049 \text{ A}}{1.732}$$

$$I_{P(SECONDARY)} = 57.765 \text{ A}$$

The amount of current flow through the primary can be calculated using the turns ratio. Because the primary has a higher voltage than the secondary, it will have a lower current. (Volts times amperes input must equal volts times amperes output.)

$$\text{Primary current} = \frac{\text{secondary current}}{\text{turns ratio}}$$

$$I_{P(PRIMARY)} = \frac{57.765 \text{ A}}{16.6}$$

$$I_{P(PRIMARY)} = 3.48 \text{ A}$$

Because all transformed values of voltage and current take place across the phases, the primary has a phase current of 3.48 A. In a wye connection, the phase current is the same as the line current:

$$I_{L(PRIMARY)} = 3.48 \text{ A}$$

The transformer connection with all calculated values is shown in *Figure 28–12*.

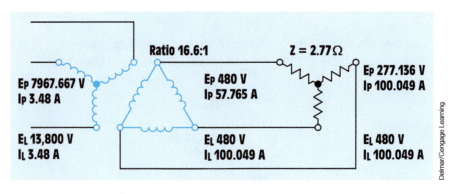

FIGURE 28–12 Example Circuit 1 with all missing values.

EXAMPLE 28-2

A three-phase transformer is connected in a delta–delta configuration *(Figure 28–13)*. The load is connected as a wye, and each phase has an impedance of 7 Ω. The primary is connected to a line voltage of 4160 V, and the secondary line voltage is 440 V. Find the following values:

$E_{P(PRIMARY)}$ — phase voltage of the primary

$E_{P(SECONDARY)}$ — phase voltage of the secondary

Ratio — turns ratio of the transformer

$E_{L(LOAD)}$ — line voltage of the load

$E_{P(LOAD)}$ — phase voltage of the load bank

$I_{P(LOAD)}$ — phase current of the load bank

$I_{L(LOAD)}$ — line current of the load

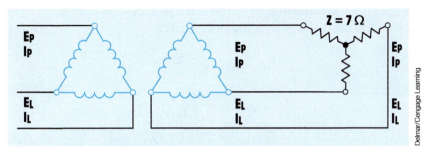

FIGURE 28–13 Example Circuit 2 three-phase transformer calculation.

$I_{L(SECONDARY)}$—secondary line current

$I_{P(SECONDARY)}$—phase current of the secondary

$I_{P(PRIMARY)}$—phase current of the primary

$I_{L(PRIMARY)}$—line current of the primary

Solution

The primary is connected as a delta. The phase voltage will be the same as the applied line voltage:

$$E_{P(PRIMARY)} = E_{L(PRIMARY)}$$
$$E_{P(PRIMARY)} = 4160 \text{ V}$$

The secondary of the transformer is connected as a delta also. Therefore, the phase voltage of the secondary will be the same as the line voltage of the secondary:

$$E_{P(SECONDARY)} = 440 \text{ V}$$

All transformer values must be calculated using phase values of voltage and current. The turns ratio can be found by dividing the phase voltage of the primary by the phase voltage of the secondary:

$$\text{Ratio} = \frac{E_{P(PRIMARY)}}{E_{P(SECONDARY)}}$$

$$\text{Ratio} = \frac{4160 \text{ V}}{440 \text{ V}}$$

$$\text{Ratio} = 9.45:1$$

The load is connected directly to the output of the secondary. The line voltage applied to the load must therefore be the same as the line voltage of the secondary:

$$E_{L(LOAD)} = 440 \text{ V}$$

The load is connected in a wye. The voltage applied across each phase will be less than the line voltage by a factor of 1.732:

$$E_{P(LOAD)} = \frac{E_{L(LOAD)}}{1.732}$$

$$E_{P(LOAD)} = \frac{440 \text{ V}}{1.732 \text{ V}}$$

$$E_{P(LOAD)} = 254.042 \text{ V}$$

The phase current of the load can be calculated using Ohm's law:

$$I_{P(LOAD)} = \frac{E_{P(LOAD)}}{Z}$$

$$I_{P(LOAD)} = \frac{254.042 \text{ V}}{7 \text{ }\Omega}$$

$$I_{P(LOAD)} = 36.292 \text{ A}$$

The amount of line current supplying a wye-connected load will be the same as the phase current of the load:

$$I_{L(LOAD)} = 36.292 \text{ A}$$

Because the secondary of the transformer is supplying current to only one load, the line current of the secondary will be the same as the line current of the load:

$$I_{L(SECONDARY)} = 36.292 \text{ A}$$

The phase current in a delta connection is less than the line current by a factor of 1.732:

$$I_{P(SECONDARY)} = \frac{I_{L(SECONDARY)}}{1.732}$$

$$I_{P(SECONDARY)} = \frac{36.292 \text{ A}}{1.732}$$

$$I_{P(SECONDARY)} = 20.954 \text{ A}$$

The phase current of the transformer primary can now be calculated using the phase current of the secondary and the turns ratio:

$$I_{P(PRIMARY)} = \frac{I_{P(SECONDARY)}}{\text{turns ratio}}$$

$$I_{P(PRIMARY)} = \frac{20.954 \text{ A}}{9.45}$$

$$I_{P(PRIMARY)} = 2.217 \text{ A}$$

In this example, the primary of the transformer is connected as a delta. The line current supplying the transformer will be higher than the phase current by a factor of 1.732:

$$I_{L(PRIMARY)} = I_{P(PRIMARY)} \times 1.732$$
$$I_{L(PRIMARY)} = 2.217 \text{ A} \times 1.732$$
$$I_{L(PRIMARY)} = 3.84 \text{ A}$$

The circuit with all calculated values is shown in *Figure 28–14*.

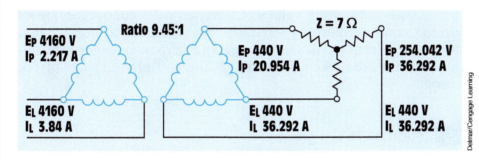

FIGURE 28–14 Example Circuit 2 with all missing values.

28–4 Open-Delta Connection

The **open-delta** transformer connection can be made with only two transformers instead of three *(Figure 28–15)*. This connection is often used when the amount of three-phase power needed is not excessive, such as in a small business. It should be noted that the output power of an open-delta connection is only 86.6% of the rated power of the two transformers. For example, assume two transformers, each having a capacity of 25 kilovolt-amperes, are connected

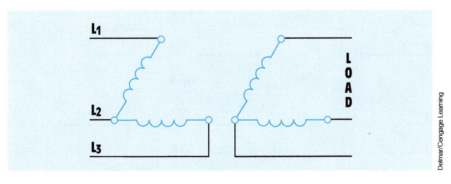

FIGURE 28–15 Open-delta connection.

in an open-delta connection. The total output power of this connection is 43.5 kilovolt-amperes (50 kVA × 0.866 = 43.3 kVA).

Another figure given for this calculation is 57.7%. This percentage assumes a closed-delta bank containing three transformers. If three 25-kilovolt-amperes transformers were connected to form a closed-delta connection, the total output power would be 75 kilovolt-amperes (3 × 25 kVA = 75 kVA). If one of these transformers were removed and the transformer bank operated as an open-delta connection, the output power would be reduced to 57.7 of its original capacity of 75 kilovolt-amperes. The output capacity of the open-delta bank is 43.3 kilovolt-amperes (75 kVA × 0.577 = 43.3 kVA).

The voltage and current values of an open-delta connection are calculated in the same manner as a standard delta–delta connection when three transformers are employed. The voltage and current rules for a delta connection must be used when determining line and phase values of voltage and current.

28–5 Single-Phase Loads

When true three-phase loads are connected to a three-phase transformer bank, there are no problems in balancing the currents and voltages of the individual phases. *Figure 28–16* illustrates this condition. In this circuit, a delta–wye three-phase transformer bank is supplying power to a wye-connected three-phase load in which the impedances of the three phases are the same. Notice that the amount of current flow in the phases is the same. This is the ideal condition and is certainly desired for all three-phase transformer loads. Although this is the ideal situation, it is not always possible to obtain a balanced load. Three-phase

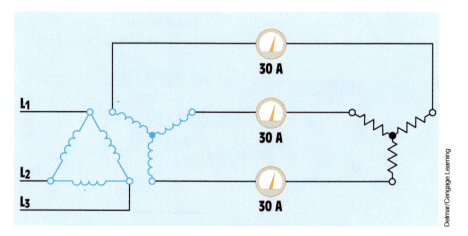

FIGURE 28–16 Three-phase transformer connected to a balanced three-phase load.

transformer connections are often used to supply **single-phase loads,** which tends to unbalance the system.

Open-Delta Connection Supplying a Single-Phase Load

The type of three-phase transformer connection used is generally determined by the amount of power needed. When a transformer bank must supply both three-phase and single-phase loads, the utility company often provides an open-delta connection with one transformer center-tapped as shown in *Figure 28–17.* In this connection, it is assumed that the amount of three-phase power needed is 20 kilovolt-amperes and the amount of single-phase power needed is 30 kilovolt-amperes. Notice that the transformer that has been center-tapped must supply power to both the three-phase and single-phase loads. Because this is an open-delta connection, the transformer bank can be loaded to only 86.6% of its full capacity when supplying a three-phase load. The rating of the three-phase transformer bank must therefore be 23 kilovolt-amperes (20 kVA/0.866 = 23 kVA). Because the rating of the two transformers can be added to obtain a total output power rating, one transformer is rated at only half the total amount of power needed, or 12 kilovolt-amperes (23 kVA/2 = 11.5 kVA). The transformer that is used to supply power to the three-phase load is only rated at 12 kilovolt-amperes. The transformer that has been center-tapped must supply power to both the single-phase and three-phase loads. Its capacity is therefore 42 kilovolt-amperes (12 kVA + 30 kVA). A 45-kilovolt-amperes transformer is used.

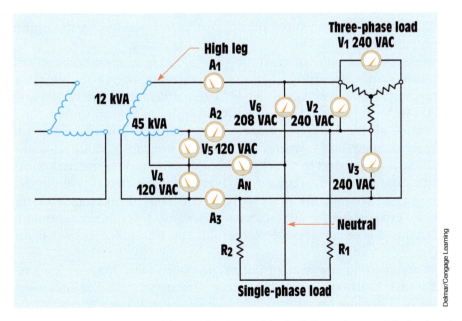

FIGURE 28–17 Three-phase open-delta transformer supplying both three-phase and single-phase loads.

Voltage Values

The connection shown in *Figure 28–17* has a line-to-line voltage of 240 volts. The three voltmeters V_1, V_2, and V_3 have all been connected across the three-phase lines and should indicate 240 volts each. Voltmeters V_4 and V_5 have been connected between the two lines of the larger transformer and its center tap. These two voltmeters will indicate a voltage of 120 volts each. Notice that it is these two lines and the center tap that are used to supply the single-phase power needed. The center tap of the larger transformer is used as a neutral conductor for the single-phase loads. Voltmeter V_6 has been connected between the center tap of the larger transformer and the line of the smaller transformer. This line is known as a **high leg** because the voltage between this line and the neutral conductor will be higher than the voltage between the neutral and either of the other two conductors. The high-leg voltage can be calculated by increasing the single-phase center-tapped voltage value by 1.732. In this case, the high-leg voltage will be 207.84 volts (120 V × 1.732 = 207.84 V). When this type of connection is employed, the *NEC* requires that the high leg be identified by connecting it to an **orange wire** or by **tagging** it at any point where a connection is made if the neutral conductor is also present.

Load Conditions

In the first load condition, it is assumed that only the three-phase load is in operation and none of the single-phase load is operating. If the three-phase load is operating at maximum capacity, Ammeters A_1, A_2, and A_3 will indicate a current flow of 48.114 amperes each [20 kVA/(240 V × 1.732) = 48.114 A]. Notice that only when the three-phase load is in operation is the current on each line balanced.

Now assume that none of the three-phase load is in operation and only the single-phase load is operating. If all the single-phase load is operating at maximum capacity, Ammeters A_2 and A_3 will each indicate a value of 125 amperes (30 kVA/240 V = 125 A). Ammeter A_1 will indicate a current flow of 0 ampere because all the load is connected between the other two lines of the transformer connection. Ammeter A_N will also indicate a value of 0 ampere. Ammeter A_N is connected in the neutral conductor, and the neutral conductor carries the sum of the unbalanced load between the two phase conductors. Another way of stating this is to say that the neutral conductor carries the difference between the two line currents. Because both these conductors are now carrying the same amount of current, the difference between them is 0 ampere.

Now assume that one side of the single-phase load, Resistor R_2, has been opened and no current flows through it. If the other line maintains a current flow of 125 amperes, the neutral conductor will have a current flow of 125 amperes also (125 A − 0 A = 125 A).

Now assume that Resistor R_2 has a value that will permit a current flow of 50 amperes on that phase. The neutral current will now be 75 amperes (125 A − 50 A = 75 A). Because the neutral conductor carries the sum of the unbalanced load, the neutral conductor never needs to be larger than the largest line conductor.

Now assume that both three-phase and single-phase loads are operating at the same time. If the three-phase load is operating at maximum capacity and the single-phase load is operating in such a manner that 125 amperes flow through Resistor R_1 and 50 amperes flow through Resistor R_2, the ammeters will indicate the following values:

$$A_1 = 48.1 \text{ A}$$
$$A_2 = 173.1 \text{ A } (48.1 \text{ A} + 125 \text{ A} = 173.1 \text{ A})$$
$$A_3 = 98.1 \text{ A } (48.1 \text{ A} + 50 \text{ A} = 98.1 \text{ A})$$
$$A_N = 75 \text{ A } (125 \text{ A} - 50 \text{ A} = 75 \text{ A})$$

Notice that the smaller of the two transformers is supplying current to only the three-phase load, but the larger transformer must supply current for both the single-phase and three-phase loads.

Although the circuit shown in *Figure 28–17* is the most common method of connecting both three-phase and single-phase loads to an open-delta transformer bank, it is possible to use the high leg to supply power to a single-phase load also. The circuit shown in *Figure 28–18* is a circuit of this type. Resistors R_1 and R_2 are connected to the lines of the transformer that has been center-tapped, and Resistor R_3 is connected to the line of the other transformer. If the line-to-line voltage is 240 volts, voltmeters V_1 and V_2 will each indicate a value of 120 volts across Resistors R_1 and R_2. Voltmeter V_3, however, will indicate that a voltage of 208 volts is applied across Resistor R_3.

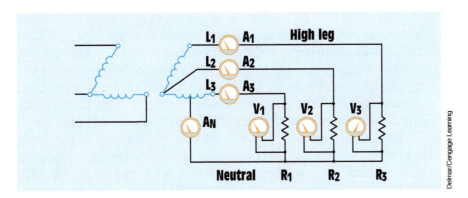

FIGURE 28–18 High leg supplies a single-phase load.

Calculating Neutral Current

The amount of current flow in the neutral conductor is still the sum of the unbalanced load between Lines L_2 and L_3, with the addition of the current flow in the high leg, L_1. To determine the amount of neutral current, use the formula

$$A_N = A_1 + (A_2 - A_3)$$

For example, assume Line L_1 has a current flow of 100 amperes, Line L_2 has a current flow of 75 amperes, and Line L_3 has a current flow of 50 amperes. The amount of current flow in the neutral conductor would be

$$A_N = A_1 + (A_2 - A_3)$$
$$A_N = 100 \text{ A} + (75 \text{ A} - 50 \text{ A})$$
$$A_N = 100 \text{ A} + 25 \text{ A}$$
$$A_N = 125 \text{ A}$$

In this circuit, it is possible for the neutral conductor to carry more current than any of the three-phase lines. This circuit is more of an example of why the *NEC* requires a high leg to be identified than it is a practical working circuit. It is rare that a high leg would be connected to the neutral conductor. This circuit is presented to illustrate the consequences that could occur and why caution should be exercised not to connect a load between the high leg and neutral.

28–6 Closed Delta with Center Tap

Another three-phase transformer configuration used to supply power to single-phase and three-phase loads is shown in *Figure 28–19*. This circuit is virtually identical to the circuit shown in *Figure 28–17* with the exception that a third transformer has been added to close the delta. Closing the delta permits more power to be supplied for the operation of three-phase loads. In this circuit, it is assumed that the three-phase load has a power requirement of 75 kilovolt-amperes and the single-phase load requires an additional 50 kilovolt-amperes. Three 25-kilovolt-ampere transformers could be used to supply the three-phase power needed (25 kVA × 3 = 75 kVA). The addition of the single-phase load, however, requires one of the transformers to be larger. This transformer must supply both the three-phase and single-phase load, which requires it to have a rating of 75 kilovolt-amperes (25 kVA + 50 kVA = 75 kVA).

In this circuit, the primary is connected in a delta configuration. Because the secondary side of the transformer bank is a delta connection, either a wye or a delta primary could have been used. This, however, is not true of all three-phase transformer connections supplying single-phase loads.

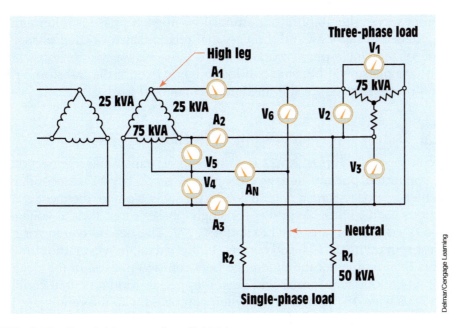

FIGURE 28–19 Closed-delta connection with high leg.

28–7 Closed Delta without Center Tap

In the circuit shown in *Figure 28–20,* the transformer bank has been connected in a wye–delta configuration. Notice that there is no transformer secondary with a center-tapped winding. In this circuit, there is no neutral conductor. The three loads have been connected directly across the three-phase lines. Because these three loads are connected directly across the lines, they form a delta-connected load. If these three loads are intended to be used as single-phase loads, they will in all likelihood have changing resistance values. The result of this connection is a three-phase delta-connected load that can be unbalanced

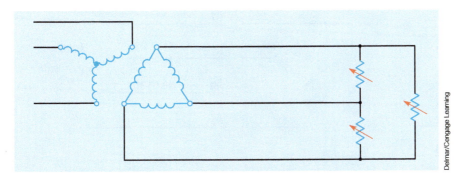

FIGURE 28–20 Single-phase loads supplied by a wye–delta transformer connection.

in different ways. The amount of current flow in each phase is determined by the impedance of the load and the vectorial relationships of each phase. Each time one of the single-phase loads is altered, the vector relationship changes also. No one phase will become overloaded, however, if the transformer bank has been properly sized for the maximum connected load.

28–8 Delta–Wye Connection with Neutral

The circuit shown in *Figure 28–21* is a three-phase transformer connection with a delta-connected primary and wye-connected secondary. The secondary has been center-tapped to form a neutral conductor. This is one of the most common connections used to provide power for single-phase loads. Typical voltages for this type of connection are 208/120 and 480/277. The neutral conductor carries the vector sum of the unbalanced current. In this circuit, however, the sum of the unbalanced current is not the difference between two phases. In the delta connection, where one transformer was center-tapped to form a neutral conductor, the two lines were 180° out of phase when compared with the center tap. In the wye connection, the lines are 120° out of phase. When all three lines are carrying the same amount of amperage, the neutral current is zero.

A wye-connected secondary with center tap can, under the right conditions, experience extreme unbalance problems. ***If this transformer connection is powered by a three-phase three-wire system, the primary winding must be connected in a delta configuration.*** If the primary is connected as a wye connection, the circuit will become exceedingly unbalanced when load is added to the circuit. Connecting the center tap of the primary to the center tap of the secondary will not solve the unbalance problem if a wye primary is used on a three-wire system.

If the incoming power is a three-phase four-wire system as shown in *Figure 28–22,* however, a wye-connected primary can be used without problem. The neutral conductor connected to the center tap of the primary prevents the

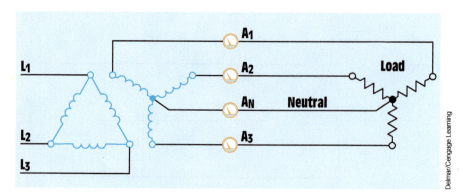

FIGURE 28–21 Three-phase four-wire connection.

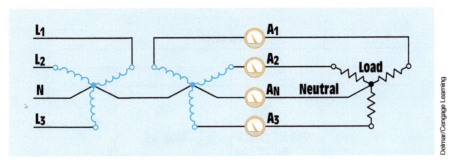

FIGURE 28–22 Neutral conductor is supplied by the incoming power.

unbalance problems. It is a common practice with this type of connection to tie the neutral conductor of both primary and secondary together as shown. When this is done, however, line isolation between the primary and secondary windings is lost.

28–9 T-Connected Transformers

Another connection involving the use of two transformers to supply three-phase power is the *T connection (Figure 28–23)*. In this connection, one transformer is generally referred to as the main transformer and the other is called the

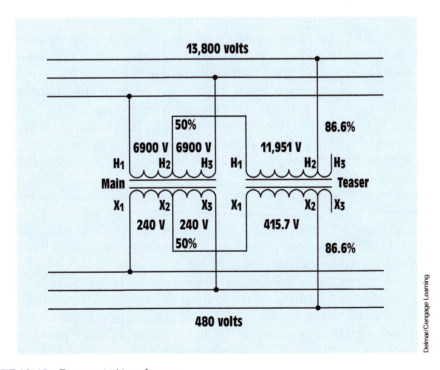

FIGURE 28–23 T-connected transformers.

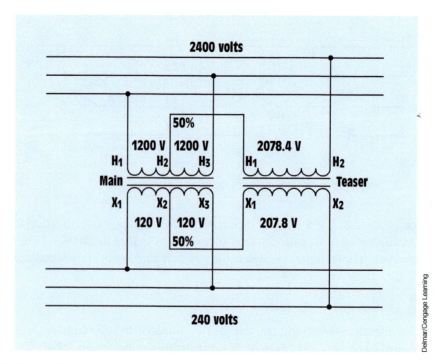

FIGURE 28–24 T-connected transformers with same voltage rating.

teaser transformer. The main transformer must contain a center or 50% tap for both the primary and secondary windings, and it is preferred that the teaser transformer contain an 86.6% voltage tap for both the primary and secondary windings. Although the 86.6% tap is preferred, the connection can be made with a teaser transformer that has the same voltage rating as the main transformer. In this instance, the teaser transformer is operated at reduced flux *(Figure 28–24)*. This connection permits two transformers to be connected T instead of open delta in the event that one transformer of a delta–delta bank should fail.

Transformers intended for use as T-connected transformers are often specially wound for the purpose, and both transformers are often contained in the same case. When making the T connection, the main transformer is connected directly across the powerline. One primary lead of the teaser transformer is connected to the center tap of the main transformer, and the 86.6% tap is connected to the powerline. The same basic connection is made for the secondary. A vector diagram illustrating the voltage relationships of the T connection is shown in *Figure 28–25*. The advantages of the T connection over the open-delta connection is that it maintains a better phase balance. This permits the T connection to be operated as a three-phase four-wire wye

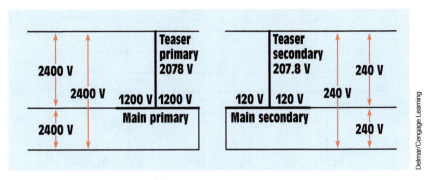

FIGURE 28–25 Voltage vector relationships of a T connection.

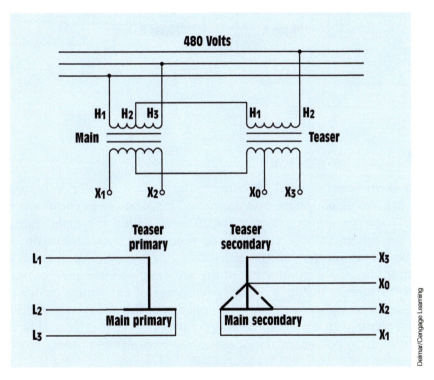

FIGURE 28–26 T-connected transformers provide a three-phase four-wire connection. X_0 is used as a center tap to the other phases.

connection like that of a wye connection with center tap *(Figure 28–26)*. When connected in this manner, the T-connected transformers can provide voltages of 480/277 or 208/120. The greatest disadvantage of the T connection is that one transformer must contain a center tap of both its primary and secondary windings.

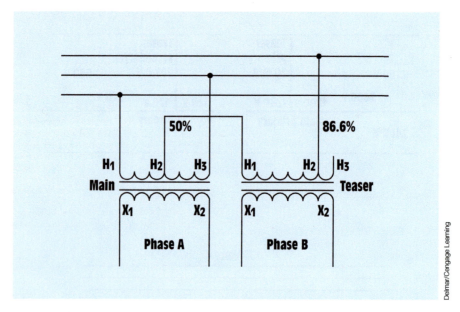

FIGURE 28–27 Scott connection.

28–10 Scott Connection

The *Scott connection* is used to convert three-phase power into two-phase power using two single-phase transformers. The Scott connection is very similar to the T connection in that one transformer, called the main transformer, must have a center, or 50% tap, and the second, or teaser transformer, must have an 86.6% tap on the primary side. The difference between the Scott and T connections lies in the connection of the secondary windings *(Figure 28–27)*. In the Scott connection, the secondary windings of each transformer provide the phases of a two-phase system. The voltages of the secondary windings are 90° out of phase with each other. The Scott connection is generally used to provide two-phase power for the operation of two-phase motors.

28–11 Zig-Zag Connection

The *zig-zag* or *interconnected-wye* transformer is primarily used for grounding purposes. It is mainly used is to establish a neutral point for the grounding of fault currents. The zig-zag connection is basically a three-phase autotransformer whose windings are divided into six equal parts *(Figure 28–28)*. In the event of a fault current, the zig-zag connection forces the current to flow equally in the three legs of the autotransformer, offering minimum impedance to the flow of fault current. A schematic diagram of the zig-zag connection is shown in *Figure 28–29*.

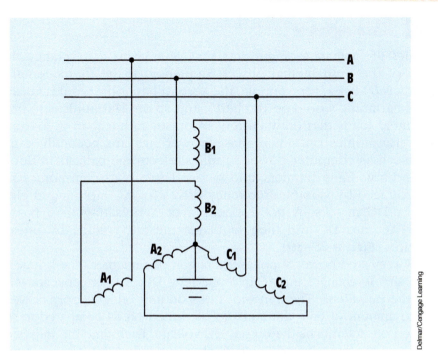

FIGURE 28–28 Zig-zag connection.

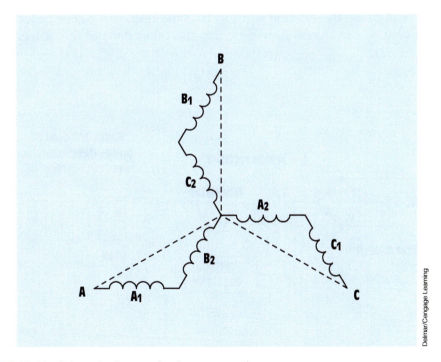

FIGURE 28–29 Schematic diagram of a zig-zag connection.

28–12 Harmonics

Harmonics are voltages or currents that operate at a frequency that is a multiple of the fundamental power frequency. If the fundamental power frequency is 60 hertz, for example, the second harmonic would be 120 hertz, the third harmonic would be 180 hertz, and so on. Harmonics are produced by nonlinear loads that draw current in pulses rather than in a continuous manner. Harmonics on single-phase powerlines are generally caused by devices such as computer power supplies, electronic ballasts in fluorescent lights, and triac light dimmers, and so on. Three-phase harmonics are generally produced by variable-frequency drives for AC motors and electronic drives for DC motors. A good example of a pulsating load is one that converts AC into DC and then regulates the DC voltage by pulse-width modulation *(Figure 28–30)*.

Many regulated power supplies operate in this manner. The bridge rectifier in *Figure 28–30* changes the AC into pulsating DC. A filter capacitor is used to smooth the pulsations. The transistor turns on and off to supply power to the load. The amount of time the transistor is turned on as compared to the time it is turned off determines the output DC voltage. Each time the transistor turns on, it causes the capacitor to begin discharging. When the transistor turns off, the capacitor will begin to charge again. Current is drawn from the AC line each time the capacitor charges. These pulsations of current produced by the charging capacitor can cause the AC sine wave to become distorted. These distorted current and voltage waveforms flow back into the other parts of the power system *(Figure 28–31)*.

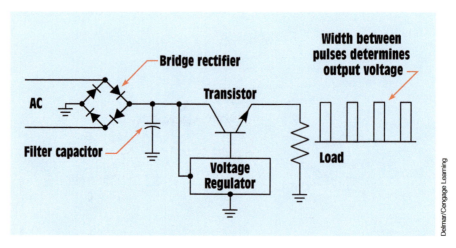

FIGURE 28–30 Pulse-width modulation regulates the output voltage by varying the time the transistor conducts as compared to the time it is turned off.

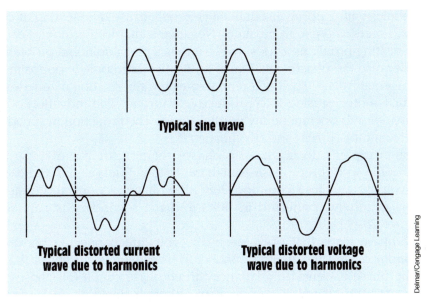

FIGURE 28–31 Harmonics cause an AC sine wave to become distorted.

Harmonic Effects

Harmonics can have very detrimental effects on electric equipment. Some common symptoms of harmonics are overheated conductors and transformers and circuit breakers that seem to trip when they should not. Harmonics are classified by name, frequency, and sequence. The name refers to whether the harmonic is the second, third, fourth, or so on of the fundamental frequency. The frequency refers to the operating frequency of the harmonic. The second harmonic operates at 120 hertz, the third at 180 hertz, the fourth at 240 hertz, and so on. The sequence refers to the phasor rotation with respect to the fundamental waveform. In an induction motor, a positive-sequence harmonic would rotate in the same direction as the fundamental frequency. A negative-sequence harmonic would rotate in the opposite direction of the fundamental frequency. A particular set of harmonics called "triplens" has a zero sequence. Triplens are the odd multiples of the third harmonic (third, ninth, fifteenth, twenty-first, etc.). A chart showing the sequence of the first nine harmonics is shown in *Table 28–1*.

Name	Fund.	2nd	3rd	4th	5th	6th	7th	8th	9th
Frequency	60	120	180	240	300	360	420	480	540
Sequence	+	±	0	+	±	0	+	±	0

TABLE 28–1. Name, Frequency, and Sequence of the First Nine Harmonics

Harmonics with a positive sequence generally cause overheating of conductors, transformers, and circuit breakers. Negative-sequence harmonics can cause the same heating problems as positive harmonics plus additional problems with motors. Because the phasor rotation of a negative harmonic is opposite that of the fundamental frequency, it will tend to weaken the rotating magnetic field of an induction motor causing it to produce less torque. The reduction of torque causes the motor to operate below normal speed. The reduction in speed results in excessive motor current and overheating.

Although triplens do not have a phasor rotation, they can cause a great deal of trouble in a three-phase four-wire system, such as a 208/120-volt or 480/277-volt system. In a common 208/120-volt wye-connected system, the primary is generally connected in delta and the secondary is connected in wye *(Figure 28–32)*.

Single-phase loads that operate on 120 volts are connected between any phase conductor and the neutral conductor. The neutral current is the vector sum of the phase currents. In a balanced three-phase circuit (all phases having equal current), the neutral current is zero. Although single-phase loads tend to cause an unbalanced condition, the vector sum of the currents generally causes the neutral conductor to carry less current than any of the phase conductors. This is true for loads that are linear and draw a continuous sine wave current. When pulsating (nonlinear) currents are connected to a three-phase four-wire system, triplen harmonic frequencies disrupt the normal phasor relationship of the phase currents and can cause the phase currents to add in the neutral conductor instead of cancel. Because the neutral conductor is not protected by a fuse or circuit breaker, there is real danger of excessive heating in the neutral conductor.

Harmonic currents are also reflected in the delta primary winding where they circulate and cause overheating. Other heating problems are caused by

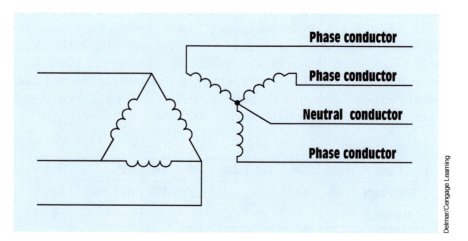

FIGURE 28–32 In a three-phase four-wire wye-connected system, the center of the wye-connected secondary is tapped to form a neutral conductor.

eddy current and hysteresis losses. Transformers are typically designed for 60-hertz operation. Higher harmonic frequencies produce greater core losses than the transformer is designed to handle. Transformers that are connected to circuits that produce harmonics must sometimes be derated or replaced with transformers that are specially designed to operate with harmonic frequencies.

Transformers are not the only electric component to be affected by harmonic currents. Emergency and standby generators can also be affected in the same way as transformers. This is especially true for standby generators used to power data-processing equipment in the event of a power failure. Some harmonic frequencies can even distort the zero crossing of the waveform produced by the generator.

Thermal-magnetic circuit breakers use a bimetallic trip mechanism that is sensitive to the heat produced by the circuit current. These circuit breakers are designed to respond to the heating effect of the true-RMS current value. If the current becomes too great, the bimetallic mechanism trips the breaker open. Harmonic currents cause a distortion of the RMS value, which can cause the breaker to trip when it should not or not to trip when it should. Thermal-magnetic circuit breakers, however, are generally better protection against harmonic currents than electronic circuit breakers. Electronic breakers sense the peak value of current. The peaks of harmonic currents are generally higher than the fundamental sine wave *(Figure 28–33)*. Although the peaks of harmonic currents are generally higher than the fundamental frequency, they can be lower. In some cases, electronic breakers may trip at low currents, and, in other cases, they may not trip at all.

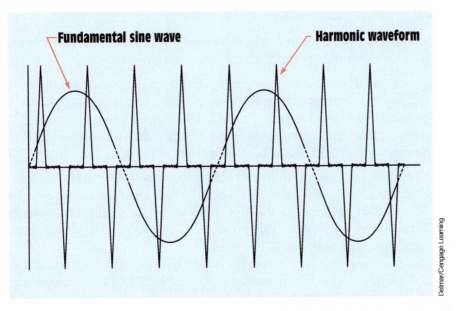

FIGURE 28–33 Harmonic waveforms generally have higher peak values than the fundamental waveform.

Triplen harmonic currents can also cause problems with neutral buss ducts and connecting lugs. A neutral buss is sized to carry the rated phase current. Because triplen harmonics can cause the neutral current to be higher than the phase current, it is possible for the neutral buss to become overloaded.

Electric panels and buss ducts are designed to carry currents that operate at 60 hertz. Harmonic currents produce magnetic fields that operate at higher frequencies. If these fields should become mechanically resonant with the panel or buss duct enclosures, the panels and buss ducts can vibrate and produce buzzing sounds at the harmonic frequency.

Telecommunications equipment is often affected by harmonic currents. Telecommunication cable is often run close to powerlines. To minimize interference, communication cables are run as far from phase conductors as possible and as close to the neutral conductor as possible. Harmonic currents in the neutral conductor induce high-frequency currents into the communication cable. These high-frequency currents can be heard as a high-pitch buzzing sound on telephone lines.

Determining Harmonic Problems on Single-Phase Systems

There are several steps to follow to determine if there is a problem with harmonics. One step is to do a survey of the equipment. This is especially important in determining if there is a problem with harmonics in a single-phase system.

1. Make an equipment check. Equipment such as personal computers, printers, and fluorescent lights with electronic ballasts are known to produce harmonics. Any piece of equipment that draws current in pulses can produce harmonics.

2. Review maintenance records to see whether there have been problems with circuit breakers tripping for no apparent reason.

3. Check transformers for overheating. If the cooling vents are unobstructed and the transformer is operating excessively hot, harmonics could be the problem. Check transformer currents with an ammeter capable of indicating a true-RMS current value. Make sure that the voltage and current ratings of the transformer have not been exceeded.

It is necessary to use an ammeter that responds to true RMS current when making this check. Some ammeters respond to the average value, not the RMS value. Meters that respond to the true-RMS value generally state this on the meter. Meters that respond to the average value are generally less expensive and do not state that they are RMS meters.

Meters that respond to the average value use a rectifier to convert the AC into DC. This value must be increased by a factor of 1.111 to change the average reading into the RMS value for a sine wave current. True-RMS responding meters calculate the heating effect of the current. The chart in *Figure 28–34*

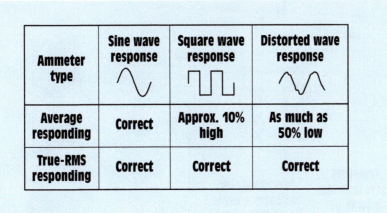

FIGURE 28–34 Comparison of average responding and true-RMS responding ammeters.

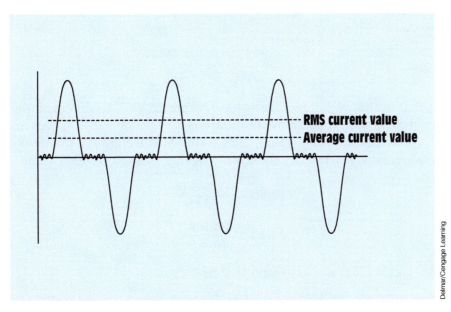

FIGURE 28–35 Average current values are generally less than the true-RMS value in a distorted waveform.

shows some of the differences between average-indicating meters and true-RMS meters. In a distorted waveform, the true-RMS value of current will no longer be *Average* × 1.111 *(Figure 28–35)*. The distorted waveform generally causes the average value to be as much as 50% less than the RMS value.

Another method of determining whether a harmonic problem exists in a single-phase system is to make two separate current checks. One check is made using an ammeter that indicates the true-RMS value and the other is made using

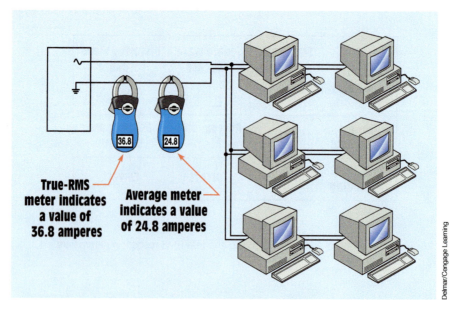

FIGURE 28–36 Determining harmonic problems using two ammeters.

a meter that indicates the average value *(Figure 28–36)*. In this example, it is assumed that the true-RMS ammeter indicates a value of 36.8 amperes and the average ammeter indicates a value of 24.8 amperes. Determine the ratio of the two measurements by dividing the average value by the true-RMS value:

$$\text{Ratio} = \frac{\text{Average}}{\text{RMS}}$$

$$\text{Ratio} = \frac{24.8 \text{ A}}{36.8 \text{ A}}$$

$$\text{Ratio} = 0.674$$

A ratio of 1 would indicate no harmonic distortion. A ratio of 0.5 would indicate extreme harmonic distortion. This method does not reveal the name or sequence of the harmonic distortion, but it does give an indication that there is a problem with harmonics. To determine the name, sequence, and amount of harmonic distortion present, a harmonic analyzer should be employed.

Determining Harmonic Problems on Three-Phase Systems

Determining whether a problem with harmonics exists in a three-phase system is similar to determining the problem in a single-phase system. Because harmonic problems in a three-phase system generally occur in a wye-connected four-wire system, this example asumes a delta-connected primary and wye-connected

Conductor	True-RMS responding ammeter	Average-responding ammeter
Phase 1	365 A	292 A
Phase 2	396 A	308 A
Phase 3	387 A	316 A
Neutral	488 A	478 A

TABLE 28–2 Three-Phase Four-Wire Wye-Connected System

secondary with a center-tapped neutral, as shown in *Figure 28–32*. To test for harmonic distortion in a three-phase four-wire system, measure all phase currents and the neutral current with both a true-RMS indicating ammeter and an average-indicating ammeter. It is assumed that the three-phase system being tested is supplied by a 200-kilovolt-ampere transformer, and the current values shown in *Table 28–2* were recorded. The current values indicate that a problem with harmonics does exist in the system. Note the higher current measurements made with the true-RMS indicating ammeter and also the fact that the neutral current is higher than any phase current.

Dealing with Harmonic Problems

After it has been determined that harmonic problems exist, something must be done to deal with the problem. It is generally not practical to remove the equipment causing the harmonic distortion, so other methods must be employed. It is a good idea to consult a power quality expert to determine the exact nature and amount of harmonic distortion present. Some general procedures for dealing with harmonics follow.

1. In a three-phase four-wire system, the 60-hertz part of the neutral current can be reduced by balancing the current on the phase conductors. If all phases have equal current flow, the neutral current would be zero.

2. If triplen harmonics are present on the neutral conductor, harmonic filters can be added at the load. These filters can help reduce the amount of harmonics on the line.

3. Pull extra neutral conductors. The ideal situation would be to use a separate neutral for each phase, instead of using a shared neutral.

4. Install a larger neutral conductor. If it is impractical to supply a separate neutral conductor for each phase, increase the size of the common neutral.

5. Derate or reduce the amount of load on the transformer. Harmonic problems generally involve overheating of the transformer. In many instances, it is necessary to derate the transformer to a point that it can handle the extra current caused by the harmonic distortion. When this is done, it is generally necessary to add a second transformer and divide the load between the two.

Determining Transformer Harmonic Derating Factor

Probably the most practical and straightforward method for determining the derating factor for a transformer is that recommended by the Computer & Business Equipment Manufacturers Association. To use this method, two ampere measurements must be made. One is the true-RMS current of the phases, and the second is the instantaneous peak phase current. The instantaneous peak current can be determined with an oscilloscope connected to a current probe or with an ammeter capable of indicating the peak value of current. Many of the digital clamp-on ammeters have the ability to indicate average, true-RMS, and peak values of current. For this example, it is assumed that peak current values are measured for the 200-kilovolt-ampere transformer discussed previously. These values are added to the previous data obtained with the true-RMS and average-indicating ammeters *(Table 28–3)*.

The formula for determining the transformer harmonic derating factor is

$$\text{THDF} = \frac{(1.414)(\text{RMS phase current})}{\text{instantaneous peak current}}$$

This formula produces a derating factor somewhere between 0 and 1.0. Because instantaneous peak value of current is equal to the RMS value $\times$ 1.414, if the current waveforms are sinusoidal (no harmonic distortion), the formula produces a derating factor of 1.0. Once the derating factor is determined, multiply the derating factor by the kilovolt-ampere capacity of the transformer. The product is the maximum load that should be placed on the transformer.

Conductor	True-RMS responding ammeter	Average-responding ammeter	Instantaneous peak current
Phase 1	365 A	292 A	716 A
Phase 2	396 A	308 A	794 A
Phase 3	387 A	316 A	737 A
Neutral	488 A	478 A	957 A

TABLE 28–3 Peak Currents Are Added to the Chart

If the phase currents are unequal, find an average value by adding the currents together and dividing by three:

$$\text{Phase (RMS)} = \frac{365 \text{ A} + 396 \text{ A} + 387 \text{ A}}{3}$$

$$\text{Phase (RMS)} = 382.7 \text{ A}$$

$$\text{Phase (Peak)} = \frac{716 \text{ A} + 794 \text{ A} + 737 \text{ A}}{3}$$

$$\text{Phase (Peak)} = 749$$

$$\text{THDF} = \frac{(1.414)(382.7 \text{ A})}{749 \text{ A}}$$

$$\text{THDF} = 0.722$$

The 200-kilovolt-ampere transformer in this example should be derated to 144.4 kilovolt-amperes (200 kVA × 0.722).

Summary

- Three-phase transformers are constructed by winding three separate transformers on the same core material.
- Single-phase transformers can be used as a three-phase transformer bank by connecting their primary and secondary windings as either wyes or deltas.
- When calculating three-phase transformer values, the rules for three-phase circuits must be followed as well as the rules for transformers.
- Phase values of voltage and current must be used when calculating the values associated with the transformer.
- The total power output of a three-phase transformer bank is the sum of the rating of the three transformers.
- An open-delta connection can be made with the use of only two transformers.
- When an open-delta connection is used, the total output power is 86.6% of the sum of the power rating of the two transformers.
- It is common practice to center-tap one of the transformers in a delta connection to provide power for single-phase loads. When this is done, the remaining phase connection becomes a high leg.
- The *NEC* requires that a high leg be identified by an orange wire or by tagging.
- The center connection of a wye is often tapped to provide a neutral conductor for three-phase loads. This produces a three-phase four-wire system. Common voltages produced by this type of connection are 208/120 and 480/277.

- Transformers should not be connected as a wye–wye unless the incoming powerline contains a neutral conductor.
- T-connected transformers provide a better phase balance than open-delta connections.
- The T connection can be used to provide a three-phase four-wire connection with only two transformers.
- The Scott connection is used to change three-phase power into two-phase power.
- The zig-zag connection is primarily used for grounding purposes.
- Harmonics are generally caused by loads that pulse the powerline.
- Harmonic distortion on single-phase lines is often caused by computer power supplies, copy machines, fax machines, and light dimmers.
- Harmonic distortion on three-phase powerlines is generally caused by variable-frequency drives and electronic DC drives.
- Harmonics can have a positive rotation, negative rotation, or no rotation.
- Positive-rotating harmonics rotate in the same direction as the fundamental frequency.
- Negative-rotating harmonics rotate in the opposite direction of the fundamental frequency.
- Triplen harmonics are the odd multiples of the third harmonic.
- Harmonic problems can generally be determined by using a true-RMS ammeter and an average-indicating ammeter, or by using a true-RMS ammeter and an ammeter that indicates the peak value.
- Triplen harmonics generally cause overheating of the neutral conductor on three-phase four-wire systems.

Review Questions

1. How many transformers are needed to make an open-delta connection?
2. Two transformers rated at 100 kVA each are connected in an open-delta connection. What is the total output power that can be supplied by this bank?
3. How does the *NEC* specify that the high leg of a four-wire delta connection be marked?
4. An open-delta three-phase transformer system has one transformer center-tapped to provide a neutral for single-phase voltages. If the voltage from line to center tap is 277 V, what is the high-leg voltage?

5. If a single-phase load is connected across the two line conductors and neutral of the transformer in Question 4 and one line has a current of 80 A and the other line has a current of 68 A, how much current is flowing in the neutral conductor?

6. A three-phase transformer connection has a delta-connected secondary, and one of the transformers has been center-tapped to form a neutral conductor. The phase-to-neutral value of the center-tapped secondary winding is 120 V. If the high leg is connected to a single-phase load, how much voltage will be applied to that load?

7. A three-phase transformer connection has a delta-connected primary and a wye-connected secondary. The center tap of the wye is used as a neutral conductor. If the line-to-line voltage is 480 V, what is the voltage between any one phase conductor and the neutral conductor?

8. A three-phase transformer bank has the secondary connected in a wye configuration. The center tap is used as a neutral conductor. If the voltage across any phase conductor and neutral is 120 V, how much voltage would be applied to a three-phase load connected to the secondary of this transformer bank?

9. A three-phase transformer bank has the primary and secondary windings connected in a wye configuration. The secondary center tap is being used as a neutral to supply single-phase loads. Will connecting the center-tap connection of the secondary to the center-tap connection of the primary permit the secondary voltage to stay in balance when a single-phase load is added to the secondary?

10. Referring to the transformer connection in Question 9, if the center tap of the primary is connected to a neutral conductor on the incoming power, will it permit the secondary voltages to be balanced when single-phase loads are added?

11. What is the frequency of the second harmonic?

12. Which of the following are considered triplen harmonics: 3rd, 6th, 9th, 12th, 15th, and 18th?

13. Would a positive-rotating harmonic or a negative-rotating harmonic be more harmful to an induction motor? Explain your answer.

14. What instrument should be used to determine what harmonics are present in a power system?

15. A 22.5-kVA single-phase transformer is tested with a true-RMS ammeter and an ammeter that indicates the peak value. The true-RMS reading is 94 A. The peak reading is 204 A. Should this transformer be derated? If so, by how much?

Practical Applications

You are working in an industrial plant. A three-phase transformer bank is connected wye–delta. The primary voltage is 12,470 V, and the secondary voltage is 480 V. The total capacity of the transformer bank is 450 kVA. One of the three transformers that form the three-phase bank develops a shorted primary winding and becomes unusable. A suggestion is made to reconnect the bank for operation as an open-delta. Can the two remaining transformers be connected open-delta? Explain your answer as to why they can or why they cannot be connected as an open-delta. If they can be reconnected open-delta, what would be the output capacity of the two remaining transformers? ■

Practical Applications

You are a journeyman electrician working in an industrial plant. You are to install transformers that are to be connected in open-delta. Transformer A must supply its share of the three-phase load. Transformer B is to be center-tapped so it can provide power to single-phase loads as well as its share of the three-phase load. The total connected three-phase load is to be 40 kVA, and the total connected single-phase load is to be 60 kVA. The transformers are to have a capacity 115% greater than the rated load. What is the minimum kVA rating of each transformer? ■

Practice Problems

Refer to the transformer shown in *Figure 28–11* and find all the missing values.

1.

Primary	Secondary	Load
E_P _____	E_P _____	E_P _____
I_P _____	I_P _____	I_P _____
E_L 4160 V	E_L 440 V	E_L _____
I_L _____	I_L _____	I_L _____
Ratio	$Z = 3.5\ \Omega$	

2.

Primary	Secondary	Load
E_P _____	E_P _____	E_P _____
I_P _____	I_P _____	I_P _____
E_L 7200 V	E_L 240 V	E_L _____
I_L _____	I_L _____	I_L _____
Ratio	$Z = 4 \, \Omega$	

Refer to the transformer connection shown in *Figure 28–37* and fill in the missing values.

3.

Primary	Secondary	Load
E_P _____	E_P _____	E_P _____
I_P _____	I_P _____	I_P _____
E_L 13,800 V	E_L 480 V	E_L _____
I_L _____	I_L _____	I_L _____
Ratio	$Z = 2.5 \, \Omega$	

4.

Primary	Secondary	Load
E_P _____	E_P _____	E_P _____
I_P _____	I_P _____	I_P _____
E_L 23,000 V	E_L 208 V	E_L _____
I_L _____	I_L _____	I_L _____
Ratio	$Z = 3 \, \Omega$	

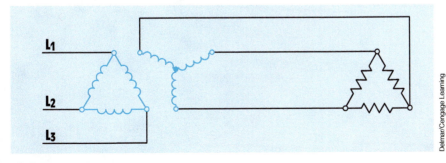

FIGURE 28–37 Practice problems circuit.

XIV AC Machines

Unit 29
Three-Phase Alternators

OUTLINE

29–1 Three-Phase Alternators
29–2 The Rotor
29–3 The Brushless Exciter
29–4 Alternator Cooling
29–5 Frequency
29–6 Output Voltage
29–7 Paralleling Alternators
29–8 Sharing the Load
29–9 Field-Discharge Protection

KEY TERMS

Alternators
Brushless exciter
Field-discharge resistor
Hydrogen
Parallel alternators
Phase rotation
Revolving-armature-type alternator
Revolving-field-type alternator
Rotor
Sliprings
Stator
Synchroscope

Why You Need to Know

Alternators produce most of the electric power in the world. Some applications for small single-phase alternators are used as portable generators for home emergency or to provide the power for portable power tools on a work site, but most alternators are three phase. This unit

- explains the principles of operation for almost all alternators regardless of size.
- discusses what determines output frequency and how output voltage is controlled.
- explains how alternators are connected in parallel to provide more power when needed.
- discusses the different types of alternators and the operation of each.
- explains how to interpret the *NEC* when determining how to connect and determine protective devices.

Objectives

After studying this unit, you should be able to

- discuss the operation of a three-phase alternator.
- explain the effect of speed of rotation on frequency.
- explain the effect of field excitation on output voltage.
- connect a three-phase alternator and make measurements using test instruments.

Preview

Most of the electric power in the world today is produced by AC generators or alternators. Electric power companies use alternators rated in gigawatts (1 gigawatt = 1,000,000,000 W) to produce the power used throughout the United States and Canada. The entire North American continent is powered by AC generators connected together in parallel. These alternators are powered by steam turbines. The turbines, called prime movers, are powered by oil, coal, natural gas, or nuclear energy. ∎

29–1 Three-Phase Alternators

Alternators operate on the same principle of electromagnetic induction as DC generators, but they have no commutator to change the AC produced in the armature into DC. There are two basic types of alternators: the revolving-armature type and the revolving-field type. Although there are some single-phase alternators that are used as portable power units for emergency home use or to operate power tools in a remote location, most alternators are three phase.

Revolving-Armature-Type Alternators

The **revolving-armature-type alternator** is the least used of the two basic types. This alternator uses an armature similar to that of a DC machine with the exception that the loops of wire are connected to **sliprings** instead of to a commutator *(Figure 29–1)*. Three separate windings are connected in either delta or wye. The armature windings are rotated inside a magnetic field *(Figure 29–2)*. Power is carried to the outside circuit via brushes riding against the sliprings. This alternator is the least used because it is very limited in the amount of output voltage and kilovolt-ampere (kVA) capacity it can develop.

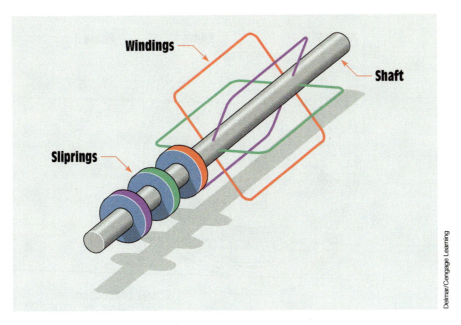

FIGURE 29–1 Basic design of a three-phase armature.

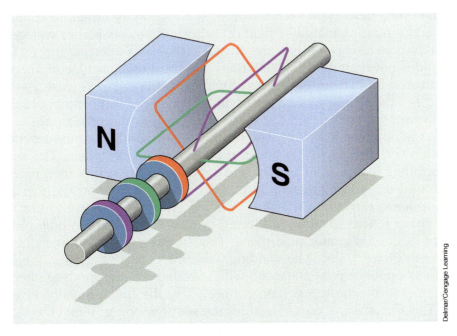

FIGURE 29–2 The armature conductors rotate inside a magnetic field.

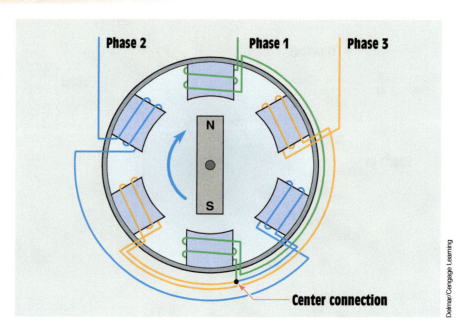

FIGURE 29–3 Basic design of a three-phase alternator.

Revolving-Field-Type Alternators

The **revolving-field-type alternator** uses a stationary armature called the **stator** and a rotating magnetic field. This design permits higher voltage and kilovolt-ampere ratings because the outside circuit is connected directly to the stator and is not routed through sliprings and brushes. This type of alternator is constructed by placing three sets of windings 120° apart *(Figure 29–3)*. In *Figure 29–3*, the winding of Phase 1 winds around the top center pole piece. It then proceeds 180° around the stator and winds around the opposite pole piece in the opposite direction. The second phase winding winds around the top pole piece directly to the left of the top center pole piece. The second phase winding is wound in an opposite direction to the first. It then proceeds 180° around the stator housing and winds around the opposite pole piece in the opposite direction. The finish end of Phase 2 connects to the finish end of Phase 1. The start end of Phase 3 winds around the top pole piece to the right of the top center pole piece. This winding is wound in a direction opposite to Phase 1 also. The winding then proceeds 180° around the stator frame to its opposite pole piece and winds around it in an opposite direction. The finish end of Phase 3 is then connected to the finish ends of Phases 1 and 2. This forms a wye connection for the stator winding. When the magnet is rotated, voltage is induced in the three windings. Because these windings are spaced 120° apart, the induced voltages are 120° out of phase with each other *(Figure 29–4)*.

FIGURE 29–4 The alternator produces three sine wave voltages 120° out of phase with each other.

The stator shown in *Figure 29–3* is drawn in a manner to aid in understanding how the three phase windings are arranged and connected. In actual practice, the stator windings are placed in a smooth cylindrical core without projecting pole pieces *(Figure 29–5)*. This design provides a better path for magnetic lines of flux and increases the efficiency of the alternator.

FIGURE 29–5 Wound stator.

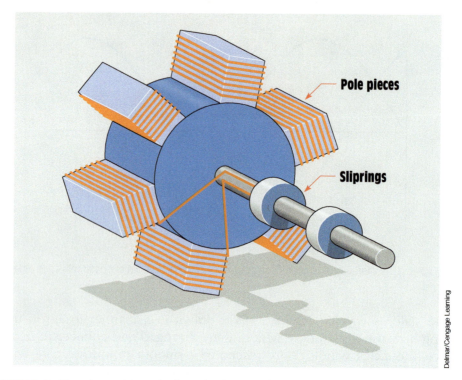

FIGURE 29–6 The rotor contains pole pieces that become electromagnets.

29–2 The Rotor

The **rotor** is the rotating member of the machine. It provides the magnetism needed to induce voltage into the stator windings. The magnets of the rotor are electromagnets and require some source of external DC to excite the alternator. This DC is known as excitation current. The alternator cannot produce an output voltage until the rotor has been excited. Some alternators use sliprings and brushes to provide the excitation current to the rotor *(Figure 29–6)*. A good example of this type of rotor can be found in the alternator of most automobiles. The DC excitation current can be varied in order to change the strength of the magnetic field. A rotor with salient (projecting) poles is shown in *Figure 29–7*.

29–3 The Brushless Exciter

Most large alternators use an exciter that contains no brushes. This is accomplished by adding a separate small alternator of the armature type on the same shaft of the rotor of the larger alternator. The armature rotates between wound

FIGURE 29–7 Rotor of the salient pole type.

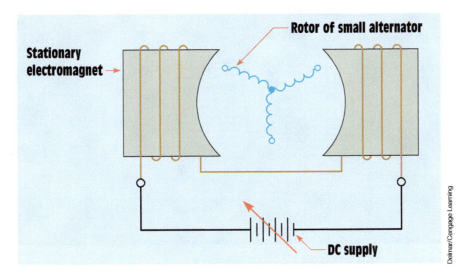

FIGURE 29–8 The brushless exciter uses stationary electromagnets.

electromagnets. The DC excitation current is connected to the wound stationary magnets *(Figure 29–8)*. The amount of voltage induced in the rotor can be varied by changing the amount of excitation current supplied to the electromagnets. The output voltage of the armature is connected to a three-phase bridge rectifier mounted on the rotor shaft *(Figure 29–9)*. The bridge rectifier converts the three-phase AC voltage produced in the armature into DC voltage

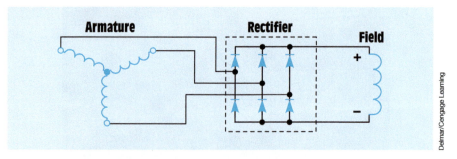

FIGURE 29–9 Basic brushless exciter circuit.

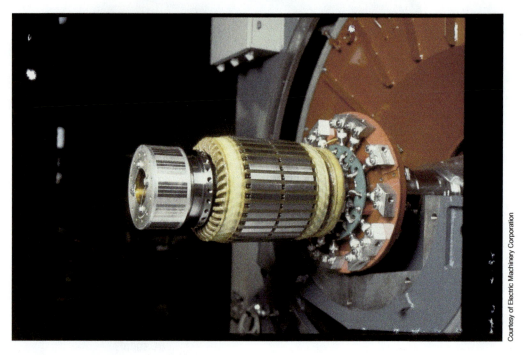

FIGURE 29–10 Brushless exciter assembly.

before it is applied to the main rotor windings. Because the armature, rectifier, and rotor winding are connected to the main rotor shaft, they all rotate together and no brushes or sliprings are needed to provide excitation current for the large alternator. A photograph of the **brushless exciter** assembly is shown in *Figure 29–10*. The field winding is placed in slots cut in the core material of the rotor *(Figure 29–11)*.

FIGURE 29–11 Two-pole rotor slotting.

29–4 Alternator Cooling

There are two main methods of cooling alternators. Alternators of small kilovolt-ampere rating are generally air-cooled. Open spaces are left in the stator windings, and slots are often provided in the core material for the passage of air. Air-cooled alternators have a fan attached to one end of the shaft that circulates air through the entire assembly.

Large-capacity alternators are often enclosed and operate in a **hydrogen** atmosphere. There are several advantages in using hydrogen. Hydrogen is less dense than air at the same pressure. The lower density reduces the windage loss of the spinning rotor. A second advantage in operating an alternator in a hydrogen atmosphere is that hydrogen has the ability to absorb and remove heat much faster than air. At a pressure of one atmosphere, hydrogen has a specific heat of approximately 3.42. The specific heat of air at a pressure of 1 atmosphere is approximately 0.238. This means that hydrogen has the ability to absorb approximately 14.37 times more heat than air. A cutaway drawing of an alternator intended to operate in a hydrogen atmosphere is shown in *Figure 29–12*.

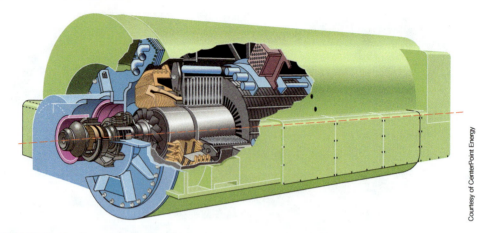

FIGURE 29–12 Two-pole, turbine-driven, hydrogen-cooled alternator.

29–5 Frequency

The output frequency of an alternator is determined by two factors:

1. the number of stator poles
2. the speed of rotation of the rotor

Because the number of stator poles is constant for a particular machine, the output frequency is controlled by adjusting the speed of the rotor. The following chart shows the speed of rotation needed to produce 60 hertz for alternators with different numbers of poles.

rpm	Stator Poles
3600	2
1800	4
1200	6
900	8

The following formula can also be used to determine the frequency when the poles and revolutions per minute (rpm) are known:

$$f = \frac{PS}{120}$$

where

f = frequency in hertz
P = number of poles per phase
S = speed in rpm
120 = a constant

EXAMPLE 29-1

What is the output frequency of an alternator that contains six poles per phase and is turning at a speed of 1000 rpm?

Solution

$$f = \frac{6 \times 1000 \text{ rpm}}{120}$$

$$f = 50 \text{ Hz}$$

29-6 Output Voltage

Three factors determine the amount of output voltage of an alternator:

1. the length of the armature or stator conductors (number of turns)
2. the strength of the magnetic field of the rotor
3. the speed of rotation of the rotor

The following formula can be used to calculate the amount of voltage induced in the stator winding:

$$E = \frac{BLV}{10^8}$$

where

10^8 = flux lines equal to 1 weber
E = induced voltage (in volts)
B = flux density in gauss
L = length of the conductor (in cm)
v = velocity (in cm/s)

One of the factors that determines the amount of induced voltage is the length of the conductor. This factor is often stated as number of turns of wire in the stator because the voltage induced in each turn adds. Increasing the number of turns of wire has the same effect as increasing the length of one conductor.

Controlling Output Voltage

The number of turns of wire in the stator cannot be changed in a particular machine without rewinding the stator, and the speed of rotation is generally maintained at a certain level to provide a constant output frequency. Therefore, the output voltage is controlled by increasing or decreasing the strength of the magnetic field of the rotor. The magnetic field strength can be controlled by controlling the DC excitation current to the rotor.

29-7 Paralleling Alternators

Because one alternator cannot produce all the power that is required, it often becomes necessary to use more than one machine. When more than one alternator is to be used, they are connected in parallel with each other. Several conditions must be met before **parallel alternators** can be used:

1. The phases must be connected in such a manner that the phase rotation of all the machines is the same.

2. Phases A, B, and C of one machine must be in sequence with Phases A, B, and C of the other machine. For example, Phase A of Alternator 1 must reach its positive peak value of voltage at the same time Phase A of Alternator 2 does *(Figure 29–13)*.

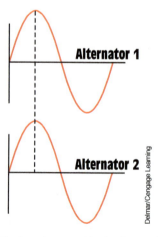

FIGURE 29–13 The voltages of both alternators must be in phase with each other.

3. The output voltage of the two alternators should be the same.
4. The frequency should be the same.

Determining Phase Rotation

The most common method of detecting when the **phase rotation** (the direction of magnetic field rotation) of one alternator is matched to the phase rotation of the other is with the use of three lights *(Figure 29–14)*. In *Figure 29–14,* the two alternators that are to be paralleled are connected together through a synchronizing switch. A set of lamps acts as a resistive load between the two machines when the switch contacts are in the open position. The voltage developed across the lamps is proportional to the difference in voltage between the two alternators. The lamps are used to indicate two conditions:

1. The lamps indicate when the phase rotation of one machine is matched to the phase rotation of the other. When both alternators are operating, both are producing a voltage. The lamps blink on and off when the phase rotation of one machine is not synchronized to the phase rotation of the other machine. If all three lamps blink on and off at the same time, or in unison, the phase rotation of Alternator 1 is correctly matched to the phase rotation of Alternator 2. If the lamps blink on and off but not in unison, the phase rotation between the two machines is not correctly matched, and two lines of Alternator 2 should be switched.

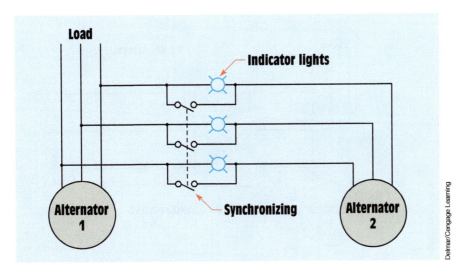

FIGURE 29–14 Determining phase rotation using indicator lights.

2. The lamps also indicate when the phase of one machine is synchronized with the phase of the other machine. If the positive peak of Alternator 1 does not occur at the same time as the positive peak of Alternator 2, there is a potential between the two machines. This permits the lamps to glow. The brightness of the lamps indicates how far out of synchronism the two machines are. When the peak voltages of the two alternators occur at the same time, there is no potential difference between them. The lamps should be off at this time. The synchronizing switch should never be closed when the lamps are glowing.

The Synchroscope

Another instrument often used for paralleling two alternators is the **synchroscope** *(Figure 29–15)*. The synchroscope measures the difference in voltage and frequency of the two alternators. The pointer of the synchroscope is free to rotate in a 360° arc. The alternator already connected to the load is considered to be the base machine. The synchroscope indicates whether the frequency of the alternator to be parallel to the base machine is fast or slow. When the voltages of the two alternators are in phase, the pointer covers the shaded area on the face of the meter. When the two alternators are synchronized, the synchronizing switch is closed.

If a synchroscope is not available, the two alternators can be paralleled using three lamps, as described earlier. If the three-lamp method is used, an AC voltmeter connected across the same phase of each machine indicates when the potential difference between the two machines is zero *(Figure 29–16)*. That is the point at which the synchronizing switch should be closed.

FIGURE 29–15 Synchroscope.

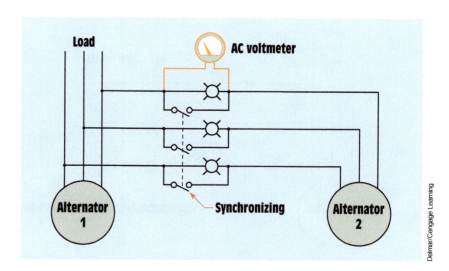

FIGURE 29–16 AC voltmeter indicates when the two alternators are in phase.

29–8 Sharing the Load

After the alternators have been paralleled, the power input to Alternator 2 must be increased to permit it to share part of the load. For example, if the alternator is being driven by a steam turbine, the power of the turbine would have to be increased. When this is done, the power to the load remains constant. The power output of the base alternator decreases, and the power output of the second alternator increases.

29–9 Field-Discharge Protection

When the DC excitation current is disconnected, the collapsing magnetic field can induce a high voltage in the rotor winding. This voltage can be high enough to arc contacts and damage the rotor winding or other circuit components. One method of preventing the induced voltage from becoming excessive is with the use of a **field-discharge resistor.** A special double-pole single-throw switch with a separate blade is used to connect the resistor to the field before the switch contacts open. When the switch is closed and DC is connected to the field, the circuit connecting the resistor to the field is open *(Figure 29–17)*. When the switch is opened, the special blade connects the resistor to the field before the main contacts open *(Figure 29–18)*.

Another method of preventing the high voltage discharge is to connect a diode in parallel with the field *(Figure 29–19)*. The diode is connected in such a manner that, when excitation current is flowing, the diode is reverse-biased and no current flows through the diode.

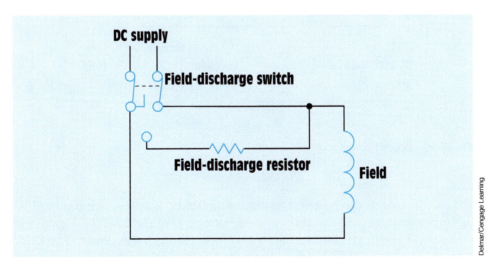

FIGURE 29–17   Switch in closed position.

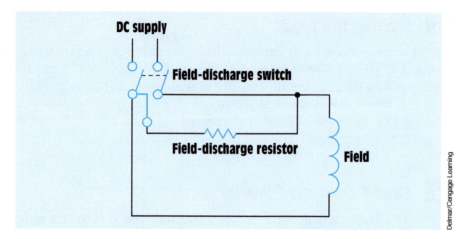

FIGURE 29–18 Switch in open position.

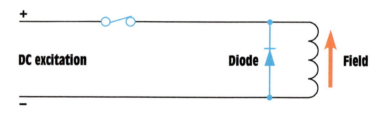

FIGURE 29–19 Normal current flow.

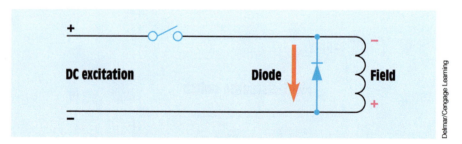

FIGURE 29–20 Induced current flow.

When the switch opens and the magnetic field collapses, the induced voltage is opposite in polarity to the applied voltage *(Figure 29–20)*. The diode is now forward-biased, permitting current to flow through the diode. The energy contained in the magnetic field is dissipated in the form of heat by the diode and field winding.

Summary

- There are two basic types of three-phase alternators: the revolving-armature type and the revolving-field type.
- The rotating-armature type is the least used because of its limited voltage and power rating.
- The rotor of the revolving-field-type alternator contains electromagnets.
- DC must be supplied to the field before the alternator can produce an output voltage.
- The DC supplied to the field is called excitation current.
- The output frequency of an alternator is determined by the number of stator poles and the speed of rotation.
- Three factors that determine the output voltage of an alternator are

    a. the length of the conductor of the armature or stator winding.

    b. the strength of the magnetic field of the rotor.

    c. the speed of the rotor.

- The output voltage is controlled by the amount of DC excitation current.
- Before two alternators can be connected in parallel, the output voltage of the two machines should be the same, the phase rotation of the machines must be the same, and the output voltages of the two machines must be in phase.
- Three lamps connected between the two alternators can be used to test for phase rotation.
- A synchroscope can be used to determine phase rotation and difference of frequency between two alternators.
- Two devices used to prevent a high voltage being induced in the rotor when the DC excitation current is stopped are a field-discharge resistor and a diode.
- Many large alternators use a brushless exciter to supply DC to the rotor winding.

Review Questions

1. What conditions must be met before two alternators can be paralleled together?
2. How can the phase rotation of one alternator be changed in relationship to the other alternator?

3. What is the function of the synchronizing lamps?

4. What is a synchroscope?

5. Assume that Alternator A is supplying power to a load and that Alternator B is to be paralleled to A. After the paralleling has been completed, what must be done to permit Alternator B to share the load with Alternator A?

6. What two factors determine the output frequency of an alternator?

7. At what speed must a six-pole alternator turn to produce 60 Hz?

8. What three factors determine the output voltage of an alternator?

9. What are sliprings used for on a revolving-field-type alternator?

10. Is the rotor excitation current AC or DC?

11. When a brushless exciter is used, what converts the AC produced in the armature winding into DC before it is supplied to the field winding?

12. What two devices are used to eliminate the induced voltage produced in the rotor when the field excitation current is stopped?

Unit 30
Three-Phase Motors

OUTLINE

30–1 Three-Phase Motors
30–2 The Rotating Magnetic Field
30–3 Connecting Dual-Voltage Three-Phase Motors
30–4 Squirrel-Cage Induction Motors
30–5 Wound-Rotor Induction Motors
30–6 Synchronous Motors
30–7 Selsyn Motors

KEY TERMS

Amortisseur winding
Code letter
Differential selsyn
Direction of rotation
Dual-voltage motors
Percent slip
Phase rotation meter
Rotating magnetic field
Rotor frequency
Selsyn motors
Single-phasing
Squirrel-cage rotor
Synchronous condenser
Synchronous speed
Wound-rotor motor

Why You Need to Know

Three-phase motors are the backbone of industry. They range in size from fractional horsepower to several thousand horsepower. It is imperative that anyone working in the electrical field have a thorough understanding of the different types of three-phase motors and their operating characteristics. This unit

- describes the three basic types of three-phase motors. Some are designed to operate on a single voltage, and others can be made to operate on two different voltages. Some motors have their stator windings connected in wye, and others are connected in delta.
- explains how stator windings are numbered and how to connect them to operate on the voltage supplied to the motor.
- describes how to calculate protective devices, different wiring methods, and operational principles.

SECTION XIV AC Machines

Objectives

After studying this unit, you should be able to

- discuss the basic operating principles of three-phase motors.
- list factors that produce a rotating magnetic field.
- list different types of three-phase motors.
- discuss the operating principles of squirrel-cage motors.
- connect dual-voltage motors for proper operation on the desired voltage.
- discuss the operation of consequent-pole motors.
- discuss the operation of wound-rotor motors.
- discuss the operation of synchronous motors.
- determine the direction of rotation of a three-phase motor using a phase rotation meter.

Preview

Three-phase motors are used throughout the United States and Canada as the prime mover for industry. These motors convert the three-phase AC into mechanical energy to operate all types of machinery. Three-phase motors are smaller and lighter and have higher efficiencies per horsepower than single-phase motors. They are extremely rugged and require very little maintenance. Many of these motors are operated 24 hours a day, 7 days a week for many years without problem. ■

30–1 Three-Phase Motors

The three basic types of three-phase motors are

1. the squirrel-cage induction motor.
2. the wound-rotor induction motor.
3. the synchronous motor.

All three motors operate on the same principle, and they all use the same basic design for the stator windings. The difference among them is the type of rotor used. Two of the three motors are induction motors and operate on the principle of electromagnetic induction in a manner similar to that of transformers. In fact, AC induction motors were patented as rotating transformers by Nikola Tesla. The stator winding of a motor is often referred to as the *motor primary*, and the rotor is referred to as the *motor secondary*.

30–2 The Rotating Magnetic Field

The operating principle for all three-phase motors is the **rotating magnetic field.** There are three factors that cause the magnetic field to rotate. These are

1. the fact that the voltages in a three-phase system are 120° out of phase with each other.

2. the fact that the three voltages change polarity at regular intervals.

3. the arrangement of the stator windings around the inside of the motor.

Figure 30–1 shows three AC sine waves 120° out of phase with each other, and the stator winding of a three-phase motor. The stator illustrates a two-pole three-phase motor. Two pole means that there are two poles per phase. AC motors do not generally have actual pole pieces as shown in *Figure 30–1,*

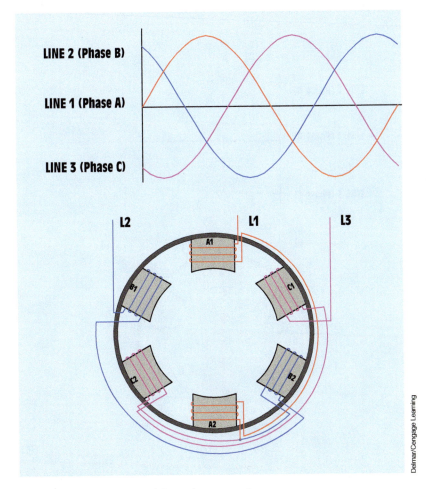

FIGURE 30–1 Three-phase stator and three sine wave voltages.

but they will be used here to aid in understanding how the rotating magnetic field is created in a three-phase motor. Notice that pole pieces A1 and A2 are located opposite each other. The same is true for poles B1 and B2 and C1 and C2. Pole pieces A1 and A2 are wound in such a manner that when current flows through the winding, they will develop opposite magnetic polarities. This is also true for poles B1 and B2 and C1 and C2. The windings of poles B1 and C1 are wound in the same direction in relation to each other, but in an opposite direction from the winding of pole A1. The start end of the winding for poles A1 and A2 is connected to Line 1, the start end of the winding for poles B1 and B2 is connected to Line 2, and the start end of the winding for poles C1 and C2 is connected to Line 3. The finish ends of all three windings are joined to form a wye connection for the stator.

To understand how the magnetic field rotates around the inside of the stator, refer to *Figure 30–2*. A dashed line labeled A has been drawn through the three sine waves of the three-phase system. This line is used to illustrate the

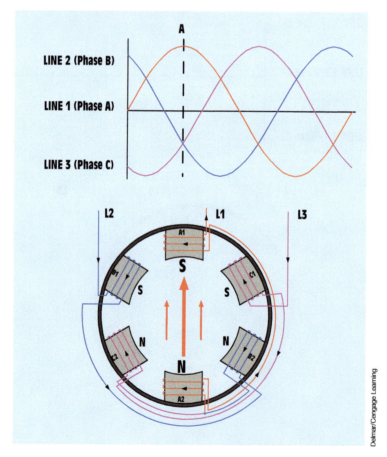

FIGURE 30–2 The magnetic field is concentrated between poles A1 and A2.

condition of the three voltages at this point in time. The arrows drawn inside the motor indicate the greatest concentration of magnetic lines of flux; the arrows are pointing in the direction that indicates magnetic lines of flux from north to south. Line 1 has reached its maximum peak voltage in the positive direction, and Lines 2 and 3 are less than maximum and in the negative direction. The magnetic field is concentrated between poles A1 and A2. Weaker lines of magnetic flux also exist between poles B1 and B2 and C1 and C2. Also note that poles A1, B1, and C1 all form a south magnetic polarity. Poles A2, B2, and C2 form a north magnetic polarity.

In *Figure 30–3,* line B is drawn at a point in time when the voltage of Line 3 is zero and the voltages of Lines 1 and 2 are less than maximum but opposite in polarity. The magnetic field is now concentrated between the pole pieces of phases A and B. Phase C has no current flow at this time and therefore no magnetic field.

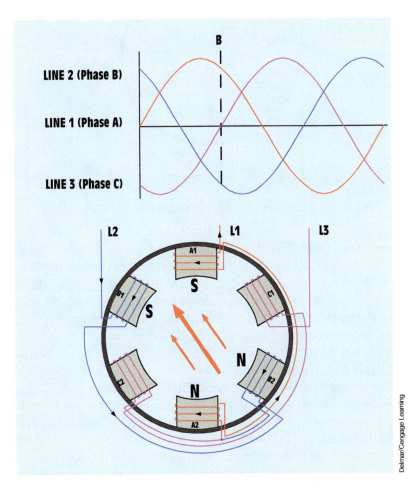

FIGURE 30–3 The magnetic field is concentrated between phases A and B.

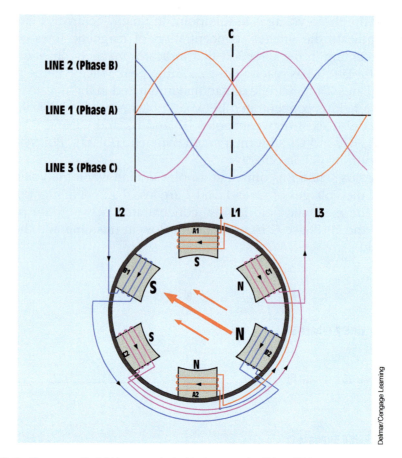

FIGURE 30–4 The magnetic field is concentrated between poles B1 and B2.

In *Figure 30–4*, line C is drawn when Line 2 has reached its maximum negative value and Lines 1 and 3 are both less than maximum and have a positive polarity. The magnetic field is concentrated between poles B1 and B2.

Line D indicates when Line 1 is zero and Lines 2 and 3 are less than maximum and opposite in polarity *(Figure 30–5)*. The magnetic field is now concentrated between the poles of phases B and C.

In *Figure 30–6,* Line E is drawn to indicate a point in time when Line 3 has reached its peak positive point and Lines 1 and 2 are less than maximum and negative. The magnetic field is now concentrated between poles C1 and C2.

Line F indicates when Line 2 is zero and Lines 1 and 3 are less than maximum and have opposite polarities *(Figure 30–7)*. The magnetic field is now concentrated between the poles of phases A and C.

In *Figure 30–8,* Line G indicates a point in time when Line 1 has reached its maximum negative value and Lines 2 and 3 are less than maximum and have a positive polarity. The magnetic field is again concentrated between poles A1 and A2. However, pole A1 has a north magnetic polarity instead of pole A2.

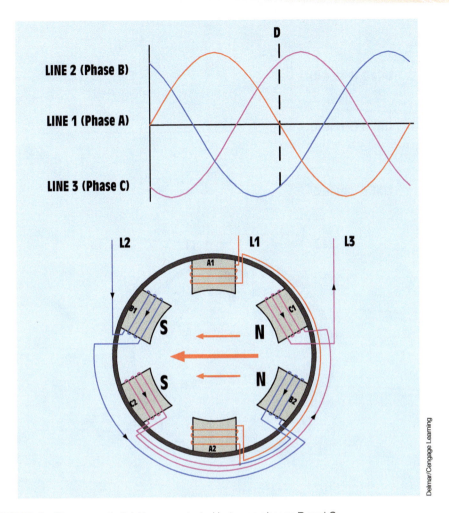

FIGURE 30–5 The magnetic field is concentrated between phases B and C.

Figure 30–9 shows the position of the magnetic at position H. The field has now rotated 270°. At the end of one complete cycle, the magnetic field completes a full 360° of rotation *(Figure 30–10)*. The speed of the rotating magnetic field is 3600 rpm in a two-pole motor connected to a 60-Hz line.

Synchronous Speed

The speed at which the magnetic field rotates is called the **synchronous speed.** Two factors that determine the synchronous speed of the rotating magnetic field are

1. the number of stator poles (per phase).
2. the frequency of the applied voltage.

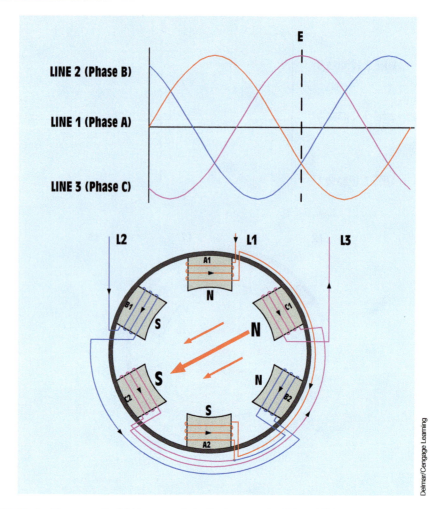

FIGURE 30–6 The magnetic field is concentrated between poles C1 and C2.

The following chart shows the synchronous speed at 60 hertz for different numbers of stator poles:

rpm	Stator poles
3600	2
1800	4
1200	6
900	8

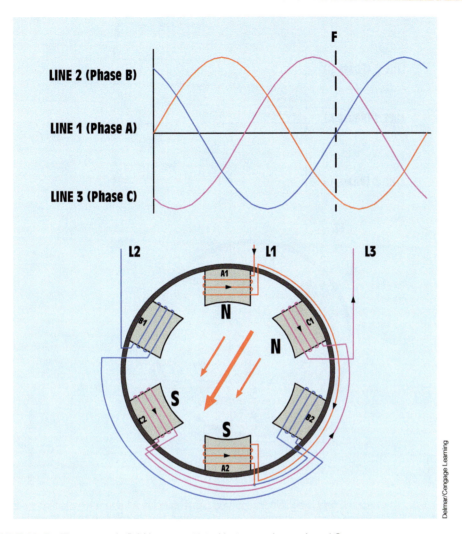

FIGURE 30–7 The magnetic field is concentrated between phases A and C.

The stator winding of a three-phase motor is shown in *Figure 30–11*. The synchronous speed can be calculated using the formula

$$S = \frac{120\,f}{P}$$

where

S = speed in rpm

f = frequency in Hz

P = number of stator poles (per phase)

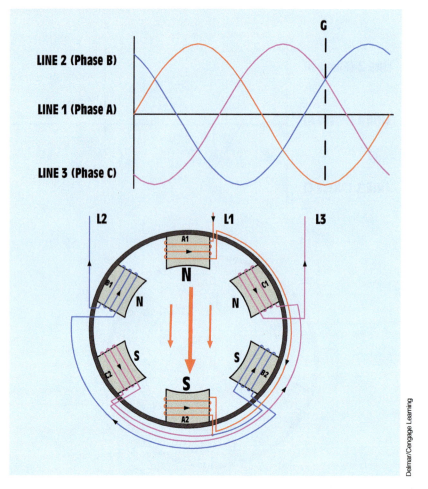

FIGURE 30–8 The magnetic field is again concentrated between poles A1 and A2. Note that the polarity of the magnetic fields has reversed. The magnetic field has rotated 180° during one half-cycle.

EXAMPLE 30–1

What is the synchronous speed of a four-pole motor connected to 50 hertz?

Solution

$$S = \frac{120 \times 50 \text{ Hz}}{4}$$

$$S = 1500 \text{ rpm}$$

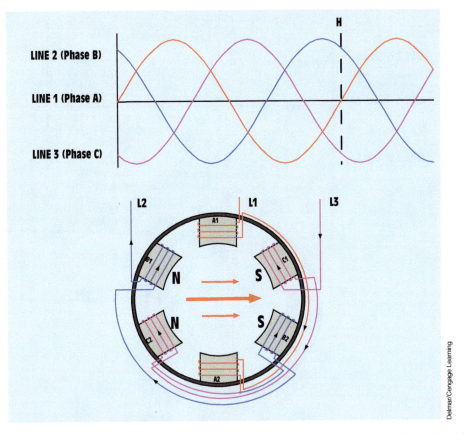

FIGURE 30–9 The magnetic field has rotated a total of 270° and is concentrated between phases B and C.

EXAMPLE 30-2

What frequency should be applied to a six-pole motor to produce a synchronous speed of 400 rpm?

Solution

First change the base formula to find frequency. Once that is done, known values can be substituted in the formula

$$f = \frac{PS}{120}$$

$$f = \frac{6 \times 400 \text{ Hz}}{120}$$

$$f = 20 \text{ Hz}$$

886 SECTION XIV AC Machines

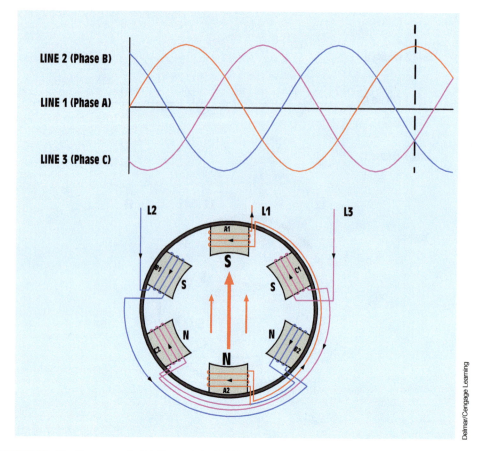

FIGURE 30–10 The magnetic field has rotated 360° after one complete cycle.

FIGURE 30–11 Stator of a three-phase motor.

Determining the Direction of Rotation for Three-Phase Motors

On many types of machinery, the direction of rotation of the motor is critical. ***The direction of rotation of any three-phase motor can be changed by reversing two of its stator leads.*** This causes the direction of the rotating magnetic field to reverse. When a motor is connected to a machine that will not be damaged when its direction of rotation is reversed, power can be momentarily applied to the motor to observe its direction of rotation. If the rotation is incorrect, any two line leads can be interchanged to reverse the motor's rotation.

When a motor is to be connected to a machine that can be damaged by incorrect rotation, however, the direction of rotation must be determined before the motor is connected to its load. The **direction of rotation** can be determined in two basic ways. One way is to make an electric connection to the motor before it is mechanically connected to the load. The direction of rotation can then be tested by momentarily applying power to the motor before it is coupled to the load.

There may be occasions when it is not practical or is very inconvenient to apply power to the motor before it is connected to the load. In such a case, a **phase rotation meter** can be used *(Figure 30–12)*. The phase rotation meter compares the phase rotation of two different three-phase connections. The meter contains six terminal leads. Three of the leads are connected to one side of the meter and are labeled MOTOR. These three motor leads are labeled A, B, or C. The LINE leads are located on the other side of the meter and are labeled A, B, or C.

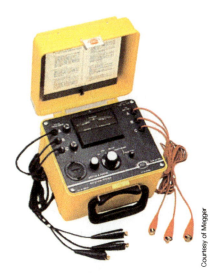

FIGURE 30–12 Phase rotation meter.

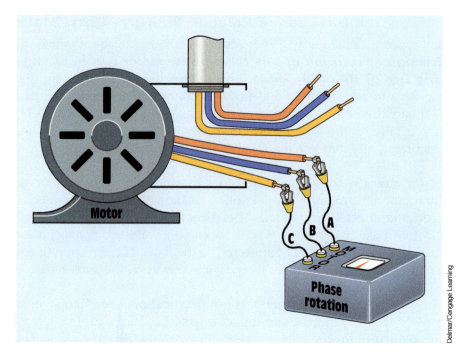

FIGURE 30–13 Connecting the phase rotation meter to the motor.

To determine the direction of rotation of the motor, first zero the meter by following the instructions provided by the manufacturer. Then set the meter selector switch to MOTOR, and connect the three MOTOR leads of the meter to the "T" leads of the motor *(Figure 30–13)*. The phase rotation meter contains a zero-center voltmeter. One side of the voltmeter is labeled INCORRECT, and the other side is labeled CORRECT. While observing the zero-center voltmeter, manually turn the motor shaft in the direction of desired rotation. The zero-center voltmeter will immediately swing in the CORRECT or INCORRECT direction. When the motor shaft stops turning, the needle may swing in the opposite direction. It is the *first* indication of the voltmeter that is to be used.

If the voltmeter needle indicates CORRECT, label the motor T leads A, B, or C to correspond with the MOTOR leads from the phase rotation meter. If the voltmeter needle indicates INCORRECT, change any two of the MOTOR leads from the phase rotation meter and again turn the motor shaft. The voltmeter needle should now indicate CORRECT. The motor T leads can now be labeled to correspond with the MOTOR leads from the phase rotation meter.

After the motor T leads have been labeled A, B, or C to correspond with the leads of the phase rotation meter, the rotation of the line supplying power to the motor must be determined. Set the selector switch on the phase rotation meter to the LINE position. After making certain the power has been turned off, connect the three LINE leads of the phase rotation meter to the incoming

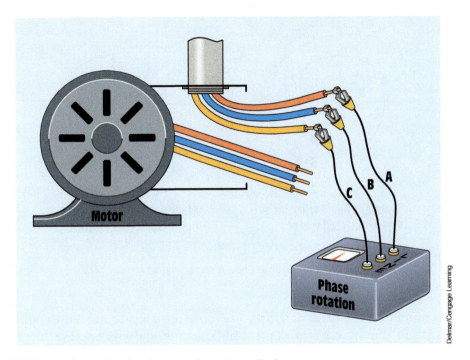

FIGURE 30–14 Connecting the phase rotation meter to the line.

powerline *(Figure 30–14)*. Turn on the power and observe the zero-center voltmeter. If the meter is pointing in the CORRECT direction, turn off the power and label the line leads A, B, or C to correspond with the LINE leads of the phase rotation meter.

If the voltmeter is pointing in the INCORRECT direction, turn off the power and change any two of the leads from the phase rotation meter. When the power is turned on, the voltmeter should point in the CORRECT direction. Turn off the power and label the line leads A, B, or C to correspond with the leads from the phase rotation meter.

Now that the motor T leads and the incoming power leads have been labeled, connect the line lead labeled A to the T lead labeled A, the line lead labeled B to the T lead labeled B, and the line lead labeled C to the T lead labeled C. When power is connected to the motor, it will operate in the proper direction.

30–3 Connecting Dual-Voltage Three-Phase Motors

Many of the three-phase motors used in industry are designed to be operated on two voltages, such as 240 volts and 480 volts. Motors of this type, called **dual-voltage motors,** contain two sets of windings per phase. Most dual-voltage motors bring out nine T leads at the terminal box. A standard method

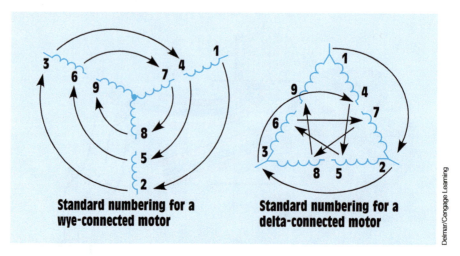

FIGURE 30–15 Standard numbering for three-phase motors.

used to number these leads is shown in *Figure 30–15*. Starting with Terminal 1, the leads are numbered in a decreasing spiral as shown. Another method of determining the proper lead numbers is to add three to each terminal. For example, starting with Lead 1, add three to one. Three plus one equals four. The phase winding that begins with 1 ends with 4. Now add three to four. Three plus four equals seven. The beginning of the second winding for phase one is seven. This method will work for the windings of all phases. If in doubt, draw a diagram of the phase windings and number them in a spiral.

High-Voltage Connections

Three-phase motors can be constructed to operate in either wye or delta. If a motor is to be connected to high voltage, the phase windings are connected in series. In *Figure 30–16,* a schematic diagram and terminal connection chart for high voltage are shown for a wye-connected motor. In *Figure 30–17,* a schematic diagram and terminal connection chart for high voltage are shown for a delta-connected motor. Notice that in both cases the windings are connected in series.

Low-Voltage Connections

When a motor is to be connected for low-voltage operation, the phase windings must be connected in parallel. *Figure 30–18* shows the basic schematic diagram for a wye-connected motor with parallel phase windings. In actual practice, however, it is not possible to make this exact connection with a nine-lead motor. The schematic shows that Terminal 4 connects to the other end of the phase winding that starts with Terminal 7. Terminal 5 connects to the other

UNIT 30 Three-Phase Motors **891**

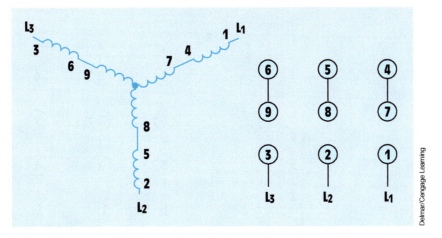

FIGURE 30–16 High-voltage wye connection.

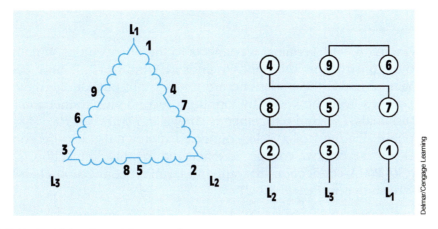

FIGURE 30–17 High-voltage delta connection.

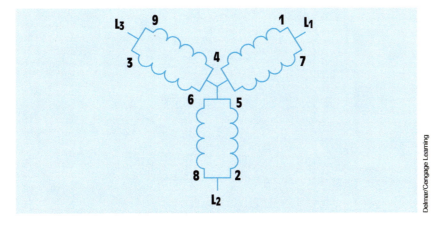

FIGURE 30–18 Stator windings connected in parallel.

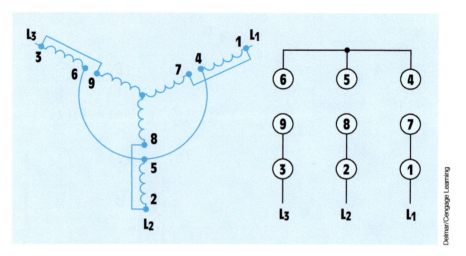

FIGURE 30–19 Low-voltage wye connection.

end of Winding 8, and Terminal 6 connects to the other end of Winding 9. In actual motor construction, the opposite ends of Windings 7, 8, and 9 are connected together inside the motor and are not brought outside the motor case. The problem is solved, however, by forming a second wye connection by connecting Terminals 4, 5, and 6 together as shown in *Figure 30–19*.

The phase windings of a delta-connected motor must also be connected in parallel for use on low voltage. A schematic for this connection is shown in *Figure 30–20*. A connection diagram and terminal connection chart for this hook-up are shown in *Figure 30–21*.

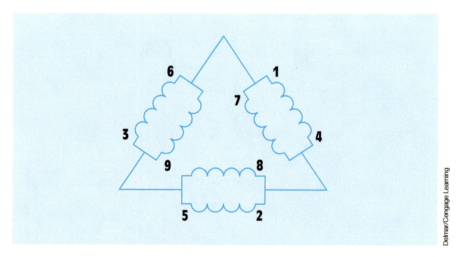

FIGURE 30–20 Parallel delta connection.

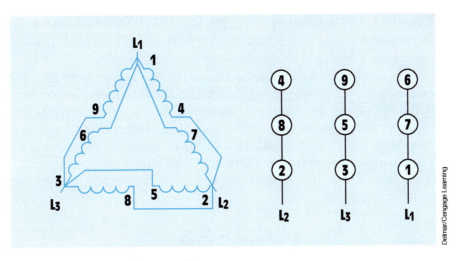

FIGURE 30-21 Low-voltage delta connection.

Some dual-voltage motors contain 12 T leads instead of 9. In this instance, the opposite ends of Terminals 7, 8, and 9 are brought out for connection. *Figure 30–22* shows the standard numbering for both delta- and wye-connected motors. Twelve leads are brought out if the motor is intended to be used for wye-delta starting. When this is the case, the motor must be designed for normal operation with its windings connected in delta. If the windings are connected in wye during starting, the starting current of the motor is reduced to one third of what it is if the motor is started as a delta.

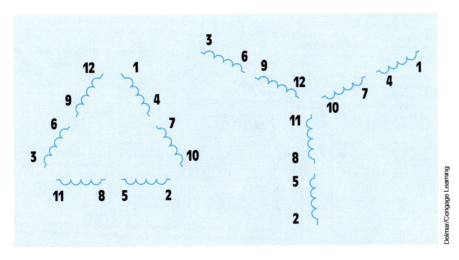

FIGURE 30-22 A 12-lead motor.

Voltage and Current Relationships for Dual-Voltage Motors

When a motor is connected to the higher voltage, the current flow will be half as much as when it is connected for low-voltage operation. The reason is that, when the windings are connected in series for high-voltage operation, the impedance is four times greater than when the windings are connected for low-voltage operation. For example, assume a dual-voltage motor is intended to operate on 480 volts or 240 volts. Also assume that during full load, the motor windings exhibit an impedance of 10 ohms each. When the winding is connected in series *(Figure 30–23)*, the impedance per phase is 20 ohms (10 Ω + 10 Ω = 20 Ω). If a voltage of 480 volts is connected to the motor, the phase voltage is

$$E_{PHASE} = \frac{E_{LINE}}{1.732}$$

$$E_{PHASE} = \frac{480}{1.732}$$

$$E_{PHASE} = 277 \text{ V}$$

The amount of current flow through the phase can be calculated using Ohm's law:

$$I = \frac{E}{Z}$$

$$I = \frac{277 \text{ V}}{20 \text{ Ω}}$$

$$I = 13.85 \text{ A}$$

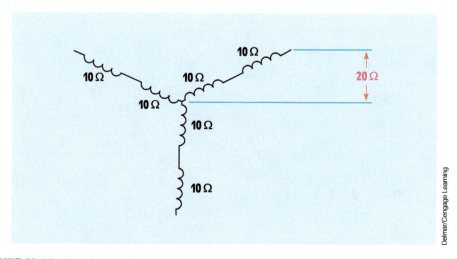

FIGURE 30–23 Impedance adds in series.

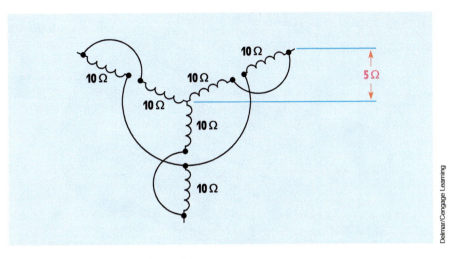

FIGURE 30–24 Impedance is less in parallel.

If the stator windings are connected in parallel, the total impedance is found by adding the reciprocals of the impedances of the windings *(Figure 30–24):*

$$Z_T = \frac{1}{\frac{1}{Z_1} + \frac{1}{Z_2}}$$

$$Z_T = 5\ \Omega$$

If a voltage of 240 volts is connected to the motor, the voltage applied across each phase is 138.6 volts (240 V/1.732 = 138.6 V). The amount of phase current can now be calculated using Ohm's law:

$$I = \frac{E}{Z}$$

$$I = \frac{138.6\ \text{V}}{5\ \Omega}$$

$$I = 27.7\ \text{A}$$

30–4 Squirrel-Cage Induction Motors

The **squirrel-cage rotor** induction motor receives its name from the type of rotor used in the motor. A squirrel-cage rotor is made by connecting bars to two end rings. If the metal laminations were removed from the rotor, the result would look very similar to a squirrel cage *(Figure 30–25)*. A squirrel cage is a cylindrical device constructed of heavy wire. A shaft placed through the center of the cage

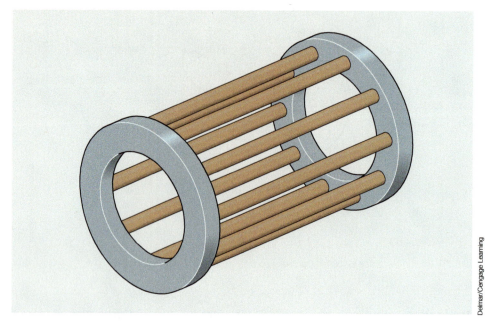

FIGURE 30–25 Basic squirrel-cage rotor without laminations.

FIGURE 30–26 Squirrel-cage rotor.

permits the cage to spin around the shaft. A squirrel cage is placed inside the cage of small pets such as squirrels and hamsters to permit them to exercise by running inside of the squirrel cage. A squirrel-cage rotor is shown in *Figure 30–26*.

Principle of Operation

The squirrel-cage motor is an induction motor. That means that the current flow in the rotor is produced by induced voltage from the rotating magnetic field of the stator. In *Figure 30–27*, a squirrel-cage rotor is shown inside the stator of a three-phase motor. It will be assumed that the motor shown in *Figure 30–27* contains four poles per phase, which produces a rotating magnetic field with a synchronous speed of 1800 rpm when the stator is connected to a 60-hertz

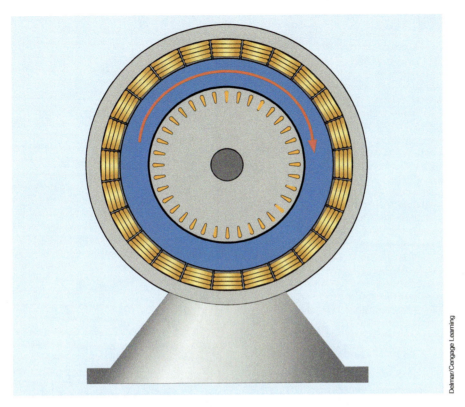

FIGURE 30–27 Voltage is induced into the rotor by the rotating magnetic field.

line. When power is first connected to the stator, the rotor is not turning. The magnetic field of the stator cuts the rotor bars at a rate of 1800 rpm. This cutting action induces a voltage into the rotor bars. This induced voltage will be the same frequency as the voltage applied to the stator. The amount of induced voltage is determined by three factors:

1. The strength of the magnetic field of the stator
2. The number of turns of wire cut by the magnetic field (in the case of a squirrel-cage rotor, this will be the number of bars in the rotor)
3. The speed of the cutting action

Because the rotor is stationary at this time, maximum voltage is induced into the rotor. The induced voltage causes current to flow through the rotor bars. As current flows through the rotor, a magnetic field is produced around each bar *(Figure 30–28)*.

The magnetic field of the rotor is attracted to the magnetic field of the stator, and the rotor begins to turn in the same direction as the rotating magnetic field.

As the speed of the rotor increases, the rotating magnetic field cuts the rotor bars at a slower rate. For example, assume the rotor has accelerated to

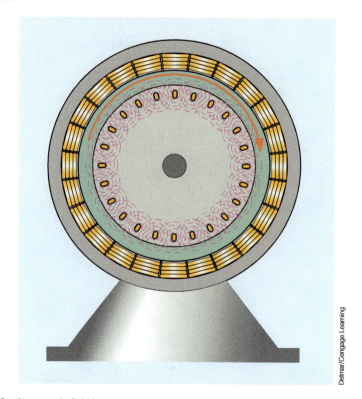

FIGURE 30–28 A magnetic field is produced around each rotor bar.

a speed of 600 rpm. The synchronous speed of the rotating magnetic field is 1800 rpm. Therefore, the rotor bars are being cut at a rate of 1200 rpm (1800 rpm − 600 rpm = 1200 rpm). Because the rotor bars are being cut at a slower rate, less voltage is induced in the rotor, reducing rotor current. When the rotor current decreases, the stator current decreases also.

As the rotor continues to accelerate, the rotating magnetic field cuts the rotor bars at a decreasing rate. This reduces the amount of induced voltage and therefore the amount of rotor current. If the motor is operating without a load, the rotor continues to accelerate until it reaches a speed close to that of the rotating magnetic field.

Torque

The amount of torque produced by an AC induction motor is determined by three factors:

1. The strength of the magnetic field of the stator

2. The strength of the magnetic field of the rotor

3. The phase angle difference between rotor and stator fields

$$T = K_T \times \varphi_S \times I_R \times \cos \theta_R$$

where

T = torque in lb-ft

K_T = torque constant

φ_S = stator flux (constant at all speeds)

I_R = rotor current

$\cos \theta_R$ = rotor power factor

Notice that one of the factors that determines the amount of torque produced by an induction motor is the strength of the magnetic field of the rotor. *An induction motor can never reach synchronous speed.* If the rotor were to turn at the same speed as the rotating magnetic field, there would be no induced voltage in the rotor and consequently no rotor current. Without rotor current, there could be no magnetic field developed by the rotor and therefore no torque or turning force. A motor operating at no load will accelerate until the torque developed is proportional to the windage and bearing friction losses.

If a load is connected to the motor, it must furnish more torque to operate the load. This causes the motor to slow down. When the motor speed decreases, the rotating magnetic field cuts the rotor bars at a faster rate. This causes more voltage to be induced in the rotor and therefore more current. The increased current flow produces a stronger magnetic field in the rotor, which causes more torque to be produced. The increased current flow in the rotor causes increased current flow in the stator. This is why motor current increases as load is added.

Another factor that determines the amount of torque developed by an induction motor is the phase angle difference between stator and rotor field flux. *Maximum torque is developed when the stator and rotor flux are in phase with each other.* Note in the preceding formula that one of the factors that determines the torque developed by an induction motor is the cosine of the rotor power factor. The cosine function reaches its maximum value of 1 when the phase angle is 0 ($\cos 0° = 1$).

Starting Characteristics

When a squirrel-cage motor is first started, it has a current draw several times greater than its normal running current. The actual amount of starting current is determined by the type of rotor bars, the horsepower rating of the motor, and the applied voltage. The type of rotor bars is indicated by the code letter found on the nameplate of a squirrel-cage motor. *Table 430.7(B)* of the *NEC* can be used to calculate the locked rotor current (starting current) of a squirrel-cage motor when the applied voltage, horsepower, and code letter are known.

EXAMPLE 30-3

An 800-hp, three-phase squirrel-cage motor is connected to 2300 V. The motor has a code letter of J. What is the starting current of this motor?

Solution

Table 430.7(B) of the *NEC* gives a value of 7.1 to 7.99 kVAs per hp as the locked-rotor current of a motor with a code letter J *(Figure 30–29)*. An average value of 7.5 is used for this calculation. The apparent power can be calculated by multiplying the 7.5 times the hp rating of the motor:

$$kVA = 7.5 \text{ kVA/hp} \times 800 \text{ hp}$$
$$kVA = 6000$$

The line current supplying the motor can now be calculated using the formula

$$I_{(LINE)} = \frac{VA}{E_{(LINE)} \times 1.732}$$

$$I_{(LINE)} = \frac{6,000,000 \text{ VA}}{2300 \text{ V} \times 1.732}$$

$$I_{(LINE)} = 1506.175 \text{ A}$$

This large starting current is caused by the fact that the rotor is not turning when power is first applied to the stator. Because the rotor is not turning, the squirrel-cage bars are cut by the rotating magnetic field at a fast rate. Remember that one of the factors that determines the amount of induced voltage is speed of the cutting action. This high induced voltage causes a large amount of current to flow in the rotor. The large current flow in the rotor causes a large amount of current flow in the stator. Because a large amount of current flows in both the stator and rotor, a strong magnetic field is established in both.

It would first appear that the starting torque of a squirrel-cage motor is high because the magnetic fields of both the stator and rotor are strong at this point. Recall that the third factor for determining the torque developed by an induction motor is the difference in phase angle between stator flux and rotor flux. Because the rotor is being cut at a high rate of speed by the rotating stator field, the bars in the squirrel-cage rotor appear to be very inductive at this point because of the high frequency of the induced voltage. This causes the phase angle difference between the induced voltage in the rotor and rotor current to

Table 430.7(B) Locked-Rotor Indicating Code Letters

Code Letter	Kilovolt-Amperes per Horsepower with Locked Rotor
A	0 – 3.14
B	3.15 – 3.54
C	3.55 – 3.99
D	4.0 – 4.49
E	4.5 – 4.99
F	5.0 – 5.59
G	5.6 – 6.29
H	6.3 – 7.09
J	7.1 – 7.99
K	8.0 – 8.99
L	9.0 – 9.99
M	10.0 – 11.19
N	11.2 – 12.49
P	12.5 – 13.99
R	14.0 – 15.99
S	16.0 – 17.99
T	18.0 – 19.99
U	20.0 – 22.39
V	22.4 and up

FIGURE 30–29 NEC Table 430.7(B). (Reprinted with permission from NFPA 70–2011)

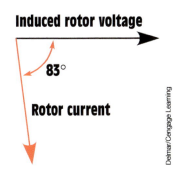

FIGURE 30–30 Rotor current is almost 90° out of phase with the induced voltage at the moment of starting.

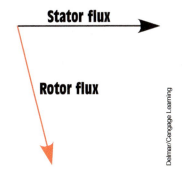

FIGURE 30–31 Rotor flux lags the stator flux by a large amount during starting.

be almost 90° out of phase with each other, producing a lagging power factor for the rotor *(Figure 30–30)*. This causes the rotor flux to lag the stator flux by a large amount; consequently, a relatively weak starting torque, per A of starting current when compared with other types of three-phase motors, is developed *(Figure 30–31)*.

Percent Slip

The speed performance of an induction motor is measured in **percent slip.** The percent slip can be determined by subtracting the synchronous speed from the speed of the rotor. For example, assume an induction motor has a synchronous speed of 1800 rpm and at full load the rotor turns at a speed of 1725 rpm. The difference between the two speeds is 75 rpm (1800 rpm − 1725 rpm = 75 rpm). The percent slip can be determined using the formula

$$\text{Percent slip} = \frac{\text{synchronous speed} - \text{rotor speed}}{\text{synchronous speed}} \times 100$$

$$\text{Percent slip} = \frac{75 \text{ rpm}}{1800 \text{ rpm}} \times 100$$

$$\text{Percent slip} = 4.16\%$$

A rotor slip of 2% to 5% is common for most squirrel-cage induction motors. The amount of slip for a particular motor is greatly affected by the type of rotor bars used in the construction of the rotor. Squirrel-cage motors are considered to be constant-speed motors because there is a small difference between noload speed and full-load speed.

Rotor Frequency

In the previous example, the rotor slips behind the rotating magnetic field by 75 rpm. This means that at full load, the bars of the rotor are being cut by magnetic lines of flux at a rate of 75 rpm. Therefore, the voltage being induced in the rotor at this point in time is at a much lower frequency than when the motor was started. The **rotor frequency** can be determined using the formula

$$f = \frac{P \times S_R}{120}$$

where

f = frequency in Hz
P = number of stator poles
S_R = rotor slip in rpm

$$f = \frac{4 \times 75 \text{ rpm}}{120}$$

$$f = 2.5 \text{ Hz}$$

Because the frequency of the current in the rotor decreases as the rotor approaches synchronous speed, the rotor bars become less inductive. The current flow through the rotor becomes limited more by the resistance of the bars and less by inductive reactance. The current flow in the rotor becomes more

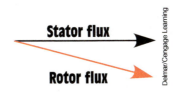

FIGURE 30–32 Rotor and stator flux become more in phase with each other as motor speed increases.

in phase with the induced voltage, which causes less phase angle shift between stator and rotor flux *(Figure 30–32)*. This is the reason that squirrel-cage motors generally have a relatively poor starting torque per ampere of starting current when compared with other types of three-phase motors but a good running torque. Although the starting torque per ampere of starting current is lower than other types of three-phase motors, the starting torque can be high because of the large amount of inrush current.

Reduced Voltage Starting

Because many squirrel-cage motors require a large amount of starting current, it is sometimes necessary to reduce the voltage during the starting period. When the voltage is reduced, the starting torque is reduced also. If the applied voltage is reduced to 50% of its normal value, the magnetic fields of both the stator and rotor are reduced to 50% of normal. The 50% reduction of the magnetic fields causes the starting torque to be reduced to 25% of normal. A chart showing a typical torque curve for a squirrel-cage motor is shown in *Figure 30–33*.

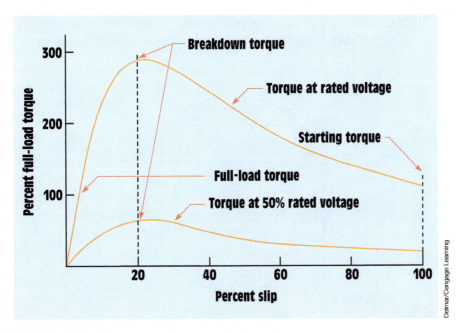

FIGURE 30–33 Typical torque curves for a squirrel-cage motor.

The torque formula given earlier can be used to show why this large reduction of torque occurs. Both the stator flux, φ_S, and the rotor current, I_R, are reduced to half their normal value. The product of these two values, torque, is reduced to one fourth. The torque varies as the square of the applied voltage for any given value of slip.

Code Letters

Squirrel-cage rotors are not all the same. Rotors are made with different types of bars. The type of rotor bars used in the construction of the rotor determines the operating characteristics of the motor. AC squirrel-cage motors are given a **code letter** on their nameplate. The code letter indicates the type of bars used in the rotor. *Figure 30–34* shows a rotor with type A bars. A type A rotor has the highest resistance of any squirrel-cage rotor. This means that the starting torque per ampere of starting current will be high because the rotor current is closer to being in phase with the induced voltage than on any other type of rotor. Also, the high resistance of the rotor bars limits the amount of current flow in the rotor when starting. This produces a low starting current for the motor. A rotor with type A bars has very poor running characteristics, however. Because the bars are resistive, a large amount of voltage will have to be induced into the rotor to produce an increase in rotor current and therefore an increase in the rotor magnetic field. This means that when load is added to the motor, the rotor must slow down a great amount to produce enough current in the rotor to increase the torque. Motors with type A rotors have the highest percent slip of any squirrel-cage motor. Motors with type A rotors are generally used in applications where starting is a problem, such as a motor that must accelerate a large flywheel from 0 rpm to its full speed. Flywheels can have a

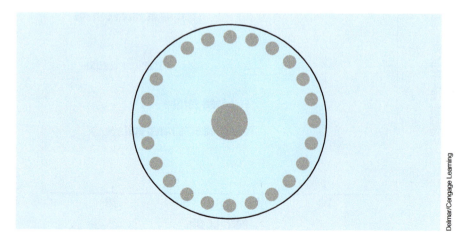

FIGURE 30–34   Type A rotor.

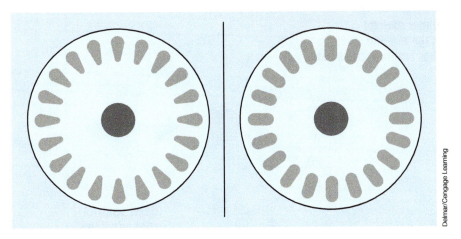

FIGURE 30–35 Type B–E rotor.   FIGURE 30–36 Type F–V rotor.

very large amount of inertia, which may require several minutes to accelerate them to their running speed when they are started.

Figure 30–35 shows a rotor with bars similar to those found in rotors with code letters B through E. These rotor bars have lower resistance than the type A rotor. Rotors of this type have fair starting torque, low starting current, and fair speed regulation.

Figure 30–36 shows a rotor with bars similar to those found in rotors with code letters F through V. This rotor has low starting torque per ampere of starting current, high starting current, and good running torque. Motors containing rotors of this type generally have very good speed regulation and low percent slip.

The Double-Squirrel-Cage Rotor

Some motors use a rotor that contains two sets of squirrel-cage windings *(Figure 30–37)*. The outer winding consists of bars with a relatively high resistance located close to the top of the iron core. Because these bars are located close to the surface, they have a relatively low reactance. The inner winding consists of bars with a large cross-sectional area, which gives them a low resistance. The inner winding is placed deeper in the core material, which causes it to have a much higher reactance.

When the double-squirrel-cage motor is started, the rotor frequency is high. Because the inner winding is inductive, its impedance will be high compared with the resistance of the outer winding. During this period of time, most of the rotor current flows through the outer winding. The resistance of the outer winding limits the current flow through the rotor, which limits the starting current to a relatively low value. Because the current is close to being in phase

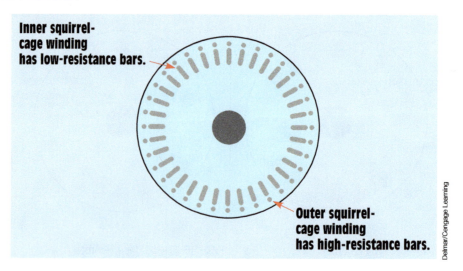

FIGURE 30–37   Double-squirrel-cage rotor.

with the induced voltage, the rotor flux and stator flux are close to being in phase with each other and a strong starting torque is developed. The starting torque of a double-squirrel-cage motor can be as high as 250% of rated full-load torque.

When the rotor reaches its full-load speed, rotor frequency decreases to 2 or 3 hertz. The inductive reactance of the inner winding has now decreased to a low value. Most of the rotor current now flows through the low-resistance inner winding. This type of motor has good running torque and excellent speed regulation.

Power Factor of a Squirrel-Cage Induction Motor

At no load, most of the current is used to magnetize the stator and rotor. Because most of the current is magnetizing current, it is inductive and lags the applied voltage by close to 90°. A very small resistive component is present, caused mostly by the resistance of the wire in the stator and the power needed to overcome bearing friction and windage loss. At no load, the motor appears to be a resistive-inductive series circuit with a large inductive component as compared with resistance *(Figure 30–38)*. A power factor of about 10% is common for a squirrel-cage motor at no load.

As load is added, electric energy is converted into mechanical energy and the in-phase component of current increases. The circuit now appears to contain more resistance than inductance *(Figure 30–39)*. This causes the phase angle between applied voltage and motor current to decrease, causing the power factor to increase. In practice, the power factor of an induction motor at full load is from about 85% to 90% lagging.

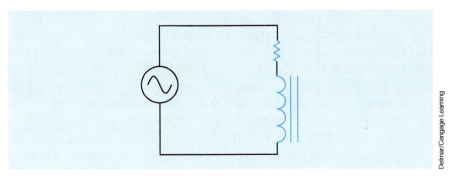

FIGURE 30–38 At no load, the motor appears to have a large amount of inductance and a very small resistance.

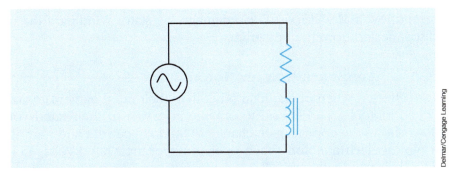

FIGURE 30–39 At full load, the resistive component of the circuit appears to be greater than the inductive component.

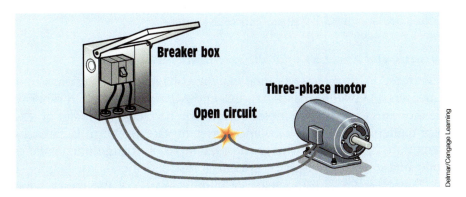

FIGURE 30–40 Single-phasing occurs when one line of a three-phase system is open.

Single-Phasing

Three lines supply power to a three-phase motor. If one of these lines should open, the motor will be connected to single-phase power *(Figure 30–40)*. This condition is known as **single-phasing.**

If the motor is not running and single-phase power is applied to the motor, the induced voltage in the rotor sets up a magnetic field in the rotor. This magnetic field opposes the magnetic field of the stator (Lenz's law). As a result, practically no torque is developed in either the clockwise or counterclockwise direction and the motor will not start. The current supplying the motor will be excessive, however, and damage to the stator windings can occur.

If the motor is operating under load at the time the single-phasing condition occurs, the rotor will continue to turn at a reduced speed. The moving bars of the rotor cut the stator field flux, which continues to induce voltage and current in the bars. Due to reduced speed, the rotor has high-reactive and low-resistive components, causing the rotor current to lag the induced voltage by almost 90°. This lagging current creates rotor fields midway between the stator poles, resulting in greatly reduced torque. The reduction in rotor speed causes high current flow and will most likely damage the stator winding if the motor is not disconnected from the powerline.

Effects of Voltage Variation on Motors

Motors are affected when operated at other than their rated nameplate voltage. NEMA rated motors are designed to operate at ±10% of their rated voltage. *Table 30–1* shows the approximate change in full-load current and starting current for typical electric motors when operated over their rated voltage (110%) and under their rated voltage (90%).

Motors are intended to operate on systems with balanced voltage (the voltage is the same between all phases). Unbalanced voltage is one of the leading causes of motor failure. Unbalanced voltage is generally caused when single-phase loads are supplied by three-phase systems.

Determining the Amount of Voltage Unbalance

The values listed in *Table 30–1* assume that the voltages across the phase conductors as measured between phases AB, BC, and AC are balanced. In other words, the table indicates the effect on motor current when voltage is greater or less than the motor nameplate rating in a balanced system. Greater harm is caused when the voltages are unbalanced. The greatest example of voltage unbalance occurs when one phase of a three-phase system is lost and the motor begins single-phasing. This causes a 173% increase of current in two of the motor windings.

Voltage Variation	Full-Load Current	Starting Current
110%	7% increase	10–12% increase
90%	11% increase	10–12% decrease

TABLE 30–1   Effects of voltage variation

If the normal full-load current of a motor is 20 amperes, the two windings still connected to power will have a current of 34.6 amperes and one winding that has lost power will have a current of 0 ampere. NEMA recommends that the unbalanced voltage not exceed ±1%. The following steps illustrate how to determine the percent of voltage unbalance in a three-phase system:

1. Take voltage measurements between all phases. In this example assume the voltage between AB = 496 volts, BC = 460 volts, and AC = 472 volts.

2. Find the average voltage.

$$\begin{array}{r}496\\460\\\underline{472}\\1428\end{array} \quad 1428 / 3 = 476 \text{ volts}$$

3. Subtract the average voltage from the voltage reading that results in the greatest difference:

$$496 - 476 = 20 \text{ volt}$$

4. Determine the percent difference:

$$\frac{100 \times \text{Greatest voltage difference}}{\text{Average voltage}}$$

$$\frac{100 \times 20}{476} = 4.2\% \text{ voltage unbalance}$$

Heat Rise

The percent of heat rise in the motor caused by the voltage unbalance is equal to twice the percent squared [2 × (percent voltage unbalance)^2]:

$$2 \times 4.2 \times 4.2 = 35.28\% \text{ temperature increase in the winding with the highest current.}$$

The Nameplate

Electric motors have nameplates that give a great deal of information about the motor. *Figure 30–41* illustrates the nameplate of a three-phase squirrel-cage induction motor. The nameplate shows that the motor is 10 horsepower, is a three-phase motor, and operates on 240 or 480 volts. The full-load running current of the motor is 28 amperes when operated on 240 volts and 14 amperes when operated on 480 volts. The motor is designed to be operated on a 60-hertz AC voltage and has a full-load speed of 1745 rpm. This speed indicates that the motor has four poles per phase. Because the full-load speed is 1745 rpm, the

Manufacturer	
HP 10	Phase 3
Volts 240/480	Amps 28/14
Hz 60	FL Speed 1745 RPM
Code J	SF 1.25
Temp 40 °C	NEMA Code B
FRAME XXXX	MODEL NO. XXXX

FIGURE 30-41 Motor nameplate.

synchronous speed would be 1800 rpm. The motor contains a type J squirrel-cage rotor and has a service factor of 1.25. The code letter indicating the type of rotor bars should not be confused with the NEMA code letter. In this example, the code letter used to determine locked-rotor current is J. The NEMA code letter is B. The *NEC* requires the NEMA code to be placed on the nameplate of squirrel-cage motors. It is used to determine fuse size when installing the motor. The service factor is used to determine the amperage rating of the overload protection for the motor. Some motors indicate a marked temperature rise in Celsius degrees instead of a service factor. The frame number indicates the type of mounting the motor has. *Figure 30-42* shows the schematic symbol used to represent a three-phase squirrel-cage motor.

Consequent-Pole Squirrel-Cage Motors

Consequent-pole squirrel-cage motors permit the synchronous speed to be changed by changing the number of stator poles. If the number of poles is

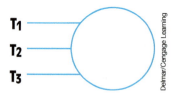

FIGURE 30-42 Schematic symbol of a three-phase squirrel-cage induction motor.

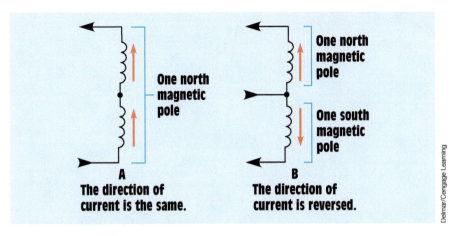

FIGURE 30–43 The number of poles can be changed by reversing the current flow through alternate poles.

doubled, the synchronous speed is reduced by one-half. A two-pole motor has a synchronous speed of 3600 rpm when operated at 60 hertz. If the number of poles is doubled to four, the synchronous speed becomes 1800 rpm. The number of stator poles can be changed by changing the direction of current flow through alternate pairs of poles.

Figure 30–43 illustrates this concept. In *Figure 30–43A,* two coils are connected in such a manner that current flows through them in the same direction. Both poles produce the same magnetic polarity and are essentially one pole. In *Figure 30–43B,* the coils have been reconnected in such a manner that current flows through them in opposite directions. The coils now produce the opposite magnetic polarities and are essentially two different poles.

Consequent-pole motors with one stator winding bring out six leads labeled T_1 through T_6. Depending on the application, the windings will be connected as a series delta or a parallel wye. If it is intended that the motor maintain the same horsepower rating for both high and low speed, the high-speed connection will be a series delta *(Figure 30–44).* The low-speed connection will be a parallel wye *(Figure 30–45).*

If it is intended that the motor maintain constant torque for both low and high speeds, the series-delta connection will provide low speed, and the parallel wye will provide high speed.

Because the speed range of a consequent-pole motor is limited to a 1:2 ratio, motors intended to operate at more than two speeds contain more than one stator winding. A consequent-pole motor with three speeds, for example, has one stator winding for one speed only and a second winding with taps. The tapped winding may provide synchronous speeds of 1800 and 900 rpm, and the separate winding may provide a speed of 1200 rpm. Consequent-pole motors with four speeds contain two separate stator windings with taps. If the

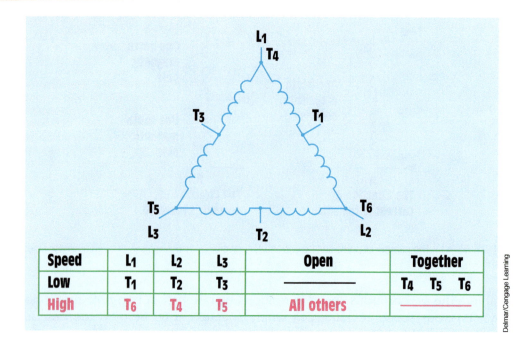

FIGURE 30–44 High-speed series-delta connection.

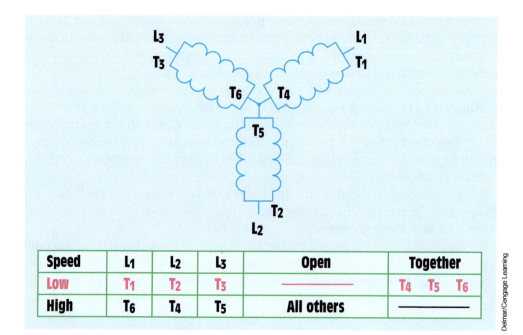

FIGURE 30–45 Low-speed parallel-wye connection.

second stator winding of the motor in this example were to be tapped, the motor would provide synchronous speeds of 1800, 1200, 900, and 600 rpm.

Motor Calculations

In the following example, output horsepower and motor efficiency are calculated. It is assumed that a 1/2-horsepower squirrel-cage motor is connected to a load. A wattmeter is connected to the motor, and the load torque measurement is calibrated in pound-inches. The motor is operating at a speed of 1725 rpm and producing a torque of 16 lb-in. The wattmeter is indicating an input power of 500 watts.

The actual amount of horsepower being produced by the motor can be calculated by using the formula

$$hp = \frac{6.28 \times rpm \times L \times P}{33,000}$$

where

hp = horsepower
6.28 = a constant
rpm = speed in revolutions per minute
L = distance in feet
P = pounds
$33,000$ = a constant

Because the formula uses feet for the distance and the torque of the motor is rated in pound-inches, L will be changed to 1/12 of a foot, or 1 inch. To simplify the calculation, the fraction 1/12 will be changed into its decimal equivalent (0.08333). To calculate the output horsepower, substitute the known values in the formula

$$hp = \frac{6.28 \times 1725 \text{ rpm} \times 0.08333 \text{ ft/in} \times 16 \text{ lb-in.}}{33,000}$$

$$hp = \frac{14,444 \text{ ft-lb/min}}{33,000}$$

$$hp = 0.438$$

One horsepower is equal to 746 watts. The output power of the motor can be calculated by multiplying the output horsepower by 746:

Power out = 746 W/hp × 0.438 hp
Power out = 326.5 W

The efficiency of the motor can be calculated by using the formula:

$$\text{Eff.} = \frac{\text{power out}}{\text{power in}} \times 100$$

$$\text{Eff.} = \frac{326.5 \text{ W}}{500 \text{ W}} \times 100$$

$$\text{Eff.} = 0.653 \times 100$$

$$\text{Eff.} = 65.3\%$$

30–5 Wound-Rotor Induction Motors

The **wound-rotor motor** induction motor is very popular in industry because of its high starting torque and low starting current. The stator winding of the wound-rotor motor is the same as the squirrel-cage motor. The difference between the two motors lies in the construction of the rotor. Recall that the squirrel-cage rotor is constructed of bars connected together at each end by a shorting ring as shown in *Figure 30–25*.

The rotor of a wound-rotor motor is constructed by winding three separate coils on the rotor 120° apart. The rotor will contain as many poles per phase as the stator winding. These coils are then connected to three sliprings located on the rotor shaft *(Figure 30–46)*. Brushes, connected to the sliprings, provide external connection to the rotor. This permits the rotor circuit to be connected to a set of resistors *(Figure 30–47)*.

The stator terminal connections are generally labeled T_1, T_2, and T_3. The rotor connections are commonly labeled M_1, M_2, and M_3. The M_2 lead is generally connected to the middle slip ring, and the M_3 lead is connected close to the rotor windings. The direction of rotation for the wound-rotor motor is reversed

FIGURE 30–46 Rotor of a wound-rotor induction motor.

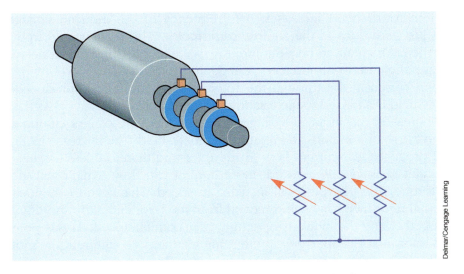

FIGURE 30–47 The rotor of a wound-rotor motor is connected to external resistors.

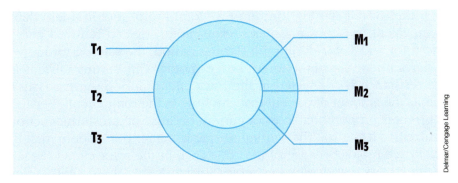

FIGURE 30–48 Schematic symbol for a wound-rotor induction motor.

by changing any two stator leads. Changing the M leads will have no effect on the direction of rotation. The schematic symbol for a wound-rotor motor is shown in *Figure 30–48*.

Principles of Operation

When power is applied to the stator winding, a rotating magnetic field is created in the motor. This magnetic field cuts through the windings of the rotor and induces a voltage into them. The amount of current flow in the rotor is determined by the amount of induced voltage and the total impedance of the rotor circuit ($I = E/Z$). The rotor impedance is a combination of inductive reactance created in the rotor windings and the external resistance. The impedance could be calculated using the formula for resistance and inductive reactance connected in series:

$$Z = \sqrt{R^2 + X_L^2}$$

As the rotor speed increases, the frequency of the induced voltage decreases just as it does in the squirrel-cage motor. The reduction in frequency causes the rotor circuit to become more resistive and less inductive, decreasing the phase angle between induced voltage and rotor current.

When current flows through the rotor, a magnetic field is produced. This magnetic field is attracted to the rotating magnetic field of the stator. As the rotor speed increases, the induced voltage decreases because of less cutting action between the rotor windings and rotating magnetic field. The decrease in induced voltage produces less current flow in the rotor and therefore less torque. If the rotor circuit resistance is reduced, more current can flow, which will increase motor torque, and the rotor will increase in speed. This action continues until all external resistance has been removed from the rotor circuit by shorting the M leads together and the motor is operating at maximum speed. At this point, the wound-rotor motor is operating in the same manner as a squirrel-cage motor.

Starting Characteristics of a Wound-Rotor Motor

Although the overall starting torque of a wound-rotor motor is less than that of an equivalent horsepower squirrel-cage motor, due to reduced current in both the rotor and stator during the starting period, the starting torque will be higher per ampere than in a squirrel-cage motor. The starting current is less because resistance is connected in the rotor circuit during starting. This resistance limits the amount of current that can flow in the rotor circuit. Because the stator current is proportional to rotor current because of transformer action, the stator current is less also. The starting torque is higher per ampere than that of a squirrel-cage motor because of the resistance in the rotor circuit. Recall that one of the factors that determines motor torque is the phase angle difference between stator flux and rotor flux. Because resistance is connected in the rotor circuit, stator and rotor flux are close to being in phase with each other producing a high starting torque for the wound-rotor induction motor. The wound-rotor motor generally exhibits a higher starting torque than an equivalent size squirrel-cage motor connected to a reduced-voltage starter.

If an attempt is made to start the motor with no circuit connected to the rotor, the motor cannot start. If no resistance is connected to the rotor circuit, there can be no current flow and consequently no magnetic field can be developed in the rotor.

Speed Control

The speed of a wound-rotor motor can be controlled by permitting resistance to remain in the rotor circuit during operation. When this is done, the rotor and stator current is limited, which reduces the strength of both magnetic fields. The reduced magnetic field strength permits the rotor to slip behind the rotating magnetic field of the stator. The resistors of speed controllers must have higher power ratings than the resistors of starters because they operate for extended periods of time.

The operating characteristics of a wound-rotor motor with the sliprings shorted are almost identical to those of a squirrel-cage motor. The percent slip, power factor, and efficiency are very similar for motors of equal horsepower rating.

30–6 Synchronous Motors

The three-phase synchronous motor has several characteristics that separate it from the other types of three-phase motors. Some of these characteristics follow.

1. The synchronous motor is not an induction motor. It does not depend on induced current in the rotor to produce a torque.
2. It will operate at a constant speed from full load to no load.
3. The synchronous motor must have DC excitation to operate.
4. It will operate at the speed of the rotating magnetic field (synchronous speed).
5. It has the ability to correct its own power factor and the power factor of other devices connected to the same line.

Rotor Construction

The synchronous motor has the same type of stator windings as the other two three-phase motors. The rotor of a synchronous motor has windings similar to the rotor of an alternator *(see Figure 30–6)*. Wound pole pieces become electromagnets when DC is applied to them. The excitation current can be applied to the rotor through two sliprings located on the rotor shaft or by a brushless exciter. The brushless exciter for a synchronous motor is the same as that used for the alternator discussed in Unit 29.

Starting a Synchronous Motor

The rotor of a synchronous motor also contains a set of squirrel-cage bars similar to those found in a type A rotor. This set of squirrel-cage bars is used to start the motor and is known as the **amortisseur winding** *(Figure 30–49)*. When power is first connected to the stator, the rotating magnetic field cuts through the squirrel-cage bars. The cutting action of the field induces a current into the squirrel-cage winding. The current flow through the amortisseur winding produces a rotor magnetic field that is attracted to the rotating magnetic field of the stator. This causes the rotor to begin turning in the direction of rotation of the stator field. When the rotor has accelerated to a speed that is close to the synchronous speed of the field, DC is connected to the rotor through the sliprings on the rotor shaft or by a brushless exciter *(Figure 30–50)*.

When DC is applied to the rotor, the windings of the rotor become electromagnets. The electromagnetic field of the rotor locks in step with the rotating magnetic field of the stator. The rotor will now turn at the same speed as the

FIGURE 30–49 Synchronous-motor rotor with amortisseur winding.

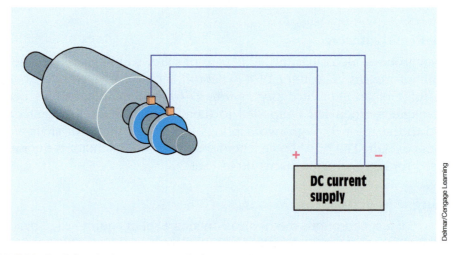

FIGURE 30–50 DC excitation current supplied through sliprings.

rotating magnetic field. When the rotor turns at the synchronous speed of the field, there is no more cutting action between the stator field and the amortisseur winding. This causes the current flow in the amortisseur winding to cease.

Notice that the synchronous motor starts as a squirrel-cage induction motor. Because the rotor uses bars that are similar to those used in a type A rotor, they have a relatively high resistance, which gives the motor good starting torque and low starting current. ***A synchronous motor must never be started with DC connected to the rotor.*** If DC is applied to the rotor, the field poles of the rotor become electromagnets. When the stator is energized, the rotating magnetic field begins turning at synchronous speed. The electromagnets are alternately attracted and repelled by the stator field. As a result, the rotor does not turn. The rotor and power supply can be damaged by high induced voltages, however.

The Field-Discharge Resistor

When the stator winding is first energized, the rotating magnetic field cuts through the rotor winding at a fast rate of speed. This causes a large amount of voltage to be induced into the winding of the rotor. To prevent this from becoming excessive, a resistor is connected across the winding. This resistor is known as the *field-discharge resistor (Figure 30–51)*. It also helps to reduce the voltage induced into the rotor by the collapsing magnetic field when the DC is disconnected from the rotor. The field-discharge resistor is connected in parallel with the rotor winding during starting. If the motor is manually started, a field-discharge switch is used to connect the excitation current to the rotor. If the motor is automatically started, a special type of relay is used to connect excitation current to the rotor and disconnect the field-discharge resistor.

Constant-Speed Operation

Although the synchronous motor starts as an induction motor, it does not operate as one. After the amortisseur winding has been used to accelerate the rotor to about 95% of the speed of the rotating magnetic field, DC is connected to the rotor and the electromagnets lock in step with the rotating field. Notice that the synchronous motor does not depend on induced voltage from the stator field to produce a magnetic field in the rotor. The magnetic field of the rotor is produced by external DC applied to the rotor. This is the reason that the synchronous motor has the ability to operate at the speed of the rotating magnetic field.

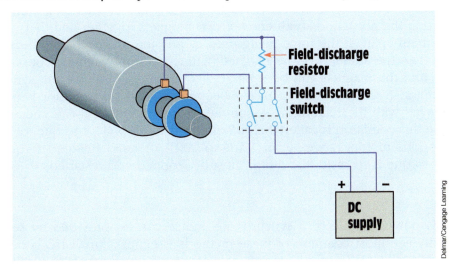

FIGURE 30–51 The field-discharge resistor is connected in parallel with the rotor winding during starting.

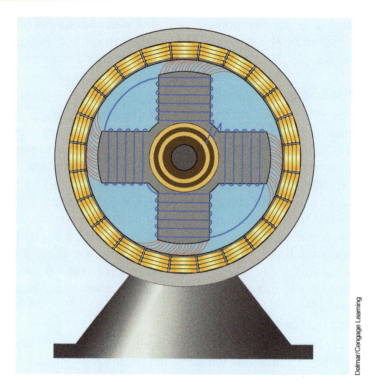

FIGURE 30–52 The magnetic field becomes stressed as load is added.

As load is added to the motor, the magnetic field of the rotor remains locked with the rotating magnetic field of the stator and the rotor continues to turn at the same speed. The added load, however, causes the magnetic fields of the rotor and stator to become stressed *(Figure 30–52)*. The action is similar to connecting the north and south ends of two magnets together and then trying to pull them apart. If the force being used to pull the magnets apart becomes greater than the strength of the magnetic attraction, the magnetic coupling is broken and the magnets can be separated. The same is true for the synchronous motor. If the load on the motor becomes too great, the rotor is pulled out of sync with the rotating magnetic field. The amount of torque necessary to cause this condition is called the *pullout torque*. The pullout torque for most synchronous motors ranges from 150% to 200% of rated full-load torque. If pullout torque is reached, the motor must be stopped and restarted.

The Power Supply

The DC power supply of a synchronous motor can be provided by several methods. The most common of these methods is either a small DC generator mounted to the shaft of the motor or an electronic power supply that converts the AC line voltage into DC voltage.

Power Factor Correction

The synchronous motor has the ability to correct its own power factor and the power factor of other devices connected to the same line. The amount of power factor correction is controlled by the amount of excitation current in the rotor. If the rotor of a synchronous motor is underexcited, the motor has a lagging power factor like a common induction motor. As rotor excitation current is increased, the synchronous motor appears to be more capacitive. When the excitation current reaches a point that the power factor of the motor is at unity or 100%, it is at the normal excitation level. At this point, the current supplying the motor drops to its lowest value.

If the excitation current is increased above the normal level, the motor has a leading power factor and appears as a capacitive load. When the rotor is overexcited, the current supplying the motor increases due to the change in power factor. The power factor at this point, however, is leading and not lagging. Because capacitance has now been added to the line, it corrects the lagging power factor of other inductive devices connected to the same line. Changes in the amount of excitation current do not affect the speed of the motor.

Interaction of the DC and AC Fields

Figure 30–53 illustrates how the magnetic flux of the AC field aids or opposes the DC field. In this example, it is assumed that the DC field is held stationary and the rotating armature is connected to the AC source. Although most synchronous motors have a stationary AC field and a rotating DC field, the principle of operation is the same. When the excitation DC is less than the amount required for normal excitation, the AC must supply some portion of the magnetizing current to aid the weak DC *(Figure 30–53A)*. This portion of magnetizing current lags the applied voltage by 90°. The current waveform shown in *Figure 30–53A* depicts only the portion of magnetizing current that is out of phase with the voltage. The remaining part of the AC is used to produce the torque necessary to operate the load. The synchronous motor has a lagging power factor at this time.

In *Figure 30–53B,* the excitation DC has been increased to the normal excitation value. All the AC is now used to produce the torque necessary to operate the load. Because the AC no longer supplies any of the magnetizing current, it is in phase with the voltage and the motor power factor is at unity or 100%. The amount of AC supplied to the motor is at its lowest value during this period.

In *Figure 30–53C,* the excitation DC is greater than that needed for normal excitation. The AC now supplies a demagnetizing component of current. The portion of AC used to demagnetize the overexcited DC field will lead the applied voltage by 90°. The current waveform shown in *Figure 30–53C* illustrates only the portion of AC used to demagnetize the DC field and does not take into account the amount of AC used to produce torque for the load. The synchronous motor now has a leading power factor.

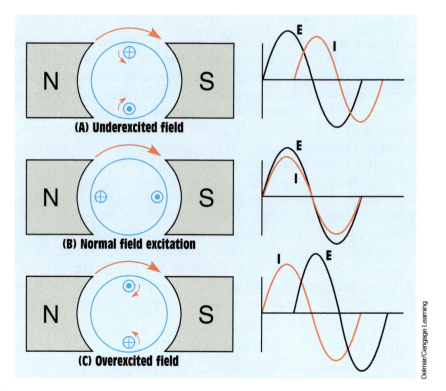

FIGURE 30–53 Field excitation in a synchronous motor.

Synchronous Motor Applications

Synchronous motors are very popular in industry, especially in the large horsepower ratings (motors up to 5000 hp are not uncommon). They have a low starting current per horsepower and a high starting torque. They operate at a constant speed from no load to full load and maintain maximum efficiency. Synchronous motors are used to operate DC generators, fans, blowers, pumps, and centrifuges. They correct their own power factor and can correct the power factor of other inductive loads connected to the same feeder *(Figure 30–54)*. Synchronous motors are sometimes operated at no load and are used for power factor correction only. When this is done, the motor is referred to as a **synchronous condenser.**

Advantages of the Synchronous Condenser

The advantage of using a synchronous condenser over a bank of capacitors for power factor correction is that the amount of correction is easily controlled. When a bank of capacitors is used for correcting power factor, capacitors must be added to or removed from the bank if a change in the amount of correction is needed. When a synchronous condenser is used, only the excitation current must be changed to cause an alteration of power factor. The schematic symbol for a synchronous motor is shown in *Figure 30–55*.

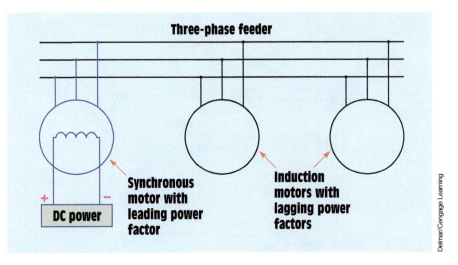

FIGURE 30–54 Synchronous motor used to correct the power factor of other motors.

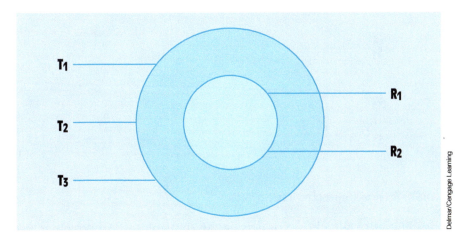

FIGURE 30–55 Schematic symbol for a synchronous motor.

30–7 Selsyn Motors

The word *selsyn* is a contraction derived from *self-synchronous*. **Selsyn motors** are used to provide position control and angular feedback information in industrial applications. Although selsyn motors are actually operated on single-phase AC, they do contain three-phase windings *(Figure 30–56)*. The schematic symbol for a selsyn motor is shown in *Figure 30–57*. This symbol is very similar to the symbol used to represent a three-phase synchronous motor. The stator windings are labeled S_1, S_2, and S_3. The rotor leads are labeled R_1 and R_2. The rotor leads are connected to the rotor winding by means of sliprings and brushes.

When selsyn motors are employed, at least two are used together. One motor is referred to as the *transmitter* and the other is called the *receiver*. It

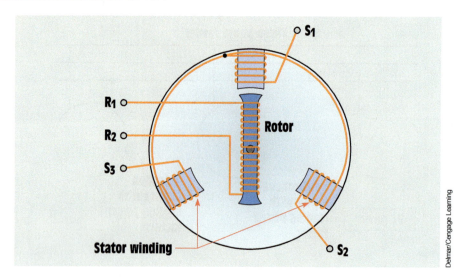

FIGURE 30–56 Selsyn motor.

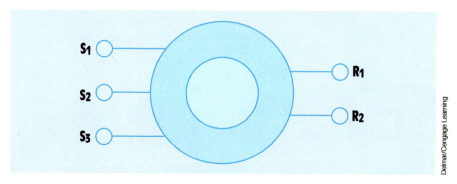

FIGURE 30–57 Schematic symbol for a selsyn motor.

makes no difference which motor acts as the transmitter and which acts as the receiver. Connection is made by connecting S_1 of the transmitter to S_1 of the receiver, S_2 of the transmitter to S_2 of the receiver, and S_3 of the transmitter to S_3 of the receiver. The rotor leads of each motor are connected to a source of single-phase AC *(Figure 30–58)*. If the stator-winding leads of the two selsyn motors are connected improperly, the receiver rotates in a direction opposite that of the transmitter. If the rotor leads are connected improperly, the rotor of the transmitter and the rotor of the receiver have an angle difference of 180°.

Selsyn Motor Operation

Selsyn motors actually operate as transformers. The rotor winding is the primary, and the stator winding is the secondary. In *Figure 30–58,* the rotor of the transmitter is in line with stator winding S_1. Because the rotor is connected to a source of AC, an alternating magnetic field exists in the rotor. This alternating

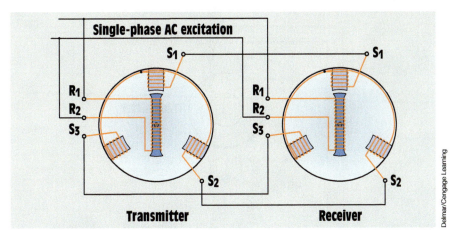

FIGURE 30-58 Connection of two selsyn motors.

magnetic field induces a voltage into the windings of the stator. Because the rotors of both motors are connected to the same source of AC, magnetic fields of identical strength and polarity exist in both motors.

Because the rotor of the transmitter is in line with stator winding S_1, maximum voltage and current are being induced in stator S_1 and less than maximum voltage and current are being induced in stator windings S_2 and S_3. Because the stator windings of the receiver are connected to the stator windings of the transmitter, the same current will flow through the receiver, producing a magnetic field in the receiver. This magnetic field attracts or repels the magnetic field of the rotor depending on the relative polarity of the two fields. When the rotor of the receiver is in the same position as the rotor of the transmitter, an equal amount of voltage is induced in the stator windings of the receiver, causing stator winding current to become zero.

If the rotor of the transmitter is turned to a different position, the magnetic field of the stator changes, resulting in a change of the magnetic field in the stator of the receiver. This causes the rotor of the receiver to rotate to a new position, where the two stator magnetic fields again cancel each other. Each time the rotor position of the transmitter is changed, the rotor of the receiver changes the same amount.

The Differential Selsyn

The **differential selsyn** is used to produce the algebraic sum of the rotation of two other selsyn units. Differential selsyns are constructed in a manner different from other selsyn motors. The differential selsyn contains three rotor windings connected in wye as well as three stator windings connected in wye. The rotor windings are brought out through three sliprings and brushes in a manner very similar to a wound-rotor induction motor. The differential selsyn

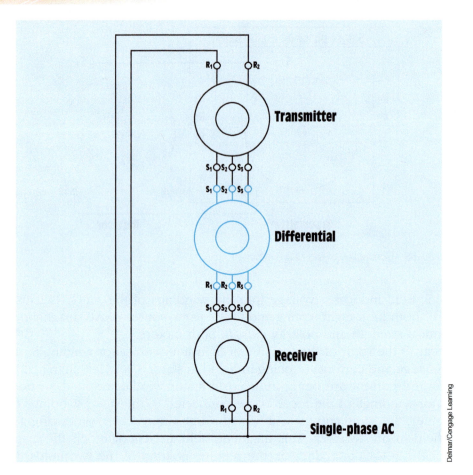

FIGURE 30–59 Differential selsyn connection.

is not connected to a source of power. Power must be provided by one of the other selsyn motors connected to it *(Figure 30–59)*.

If any one of the selsyn units is held in place and a second unit is turned, the third turns by the same amount. If any two of the selsyn units are turned at the same time, the third turns an amount equal to the sum of the angle of rotation of the other two.

Summary

- Three basic types of three-phase motors are
    a. squirrel-cage induction motor.
    b. wound-rotor induction motor.
    c. synchronous motor.
- All three-phase motors operate on the principle of a rotating magnetic field.

- Three factors that cause a magnetic field to rotate are
  a. the fact that the voltages of a three-phase system are 120° out of phase with each other.
  b. the fact that voltages change polarity at regular intervals.
  c. the arrangement of the stator windings.
- The speed of the rotating magnetic field is called the synchronous speed.
- Two factors that determine the synchronous speed are
  a. number of stator poles per phase.
  b. frequency of the applied voltage.
- The direction of rotation of any three-phase motor can be changed by reversing the connection of any two stator leads.
- The direction of rotation of a three-phase motor can be determined with a phase rotation meter before power is applied to the motor.
- Dual-voltage motors have 9 or 12 leads brought out at the terminal connection box.
- Dual-voltage motors intended for high-voltage connection have their phase windings connected in series.
- Dual-voltage motors intended for low-voltage connection have their phase windings connected in parallel.
- Motors that bring out 12 leads are generally intended for wye–delta starting.
- Three factors that determine the torque produced by an induction motor are
  a. the strength of the magnetic field of the stator.
  b. the strength of the magnetic field of the rotor.
  c. the phase angle difference between rotor and stator flux.
- Maximum torque is developed when stator and rotor flux are in phase with each other.
- The code letter on the nameplate of a squirrel-cage motor indicates the type of rotor bars used in the construction of the rotor.
- The type A rotor has the lowest starting current, highest starting torque, and poorest speed regulation of any type squirrel-cage rotor.
- The double-squirrel-cage rotor contains two sets of squirrel-cage windings in the same rotor.
- Consequent-pole squirrel-cage motors change speed by changing the number of stator poles.

- Wound-rotor induction motors have wound rotors that contain three-phase windings.
- Wound-rotor motors have three sliprings on the rotor shaft to provide external connection to the rotor.
- Wound-rotor motors have higher starting torque and lower starting current than squirrel-cage motors of the same horsepower.
- The speed of a wound-rotor motor can be controlled by permitting resistance to remain in the rotor circuit during operation.
- Synchronous motors operate at synchronous speed.
- Synchronous motors operate at a constant speed from no load to full load.
- When load is connected to a synchronous motor, stress develops between the magnetic fields of the rotor and stator.
- Synchronous motors must have DC excitation from an external source.
- DC excitation is provided to some synchronous motors through two sliprings located on the rotor shaft, and other motors use a brushless exciter.
- Synchronous motors have the ability to produce a leading power factor by overexcitation of the DC supplied to the rotor.
- Synchronous motors have a set of type A squirrel-cage bars used for starting. This squirrel-cage winding is called the amortisseur winding.
- A field-discharge resistor is connected across the rotor winding during starting to prevent high voltage in the rotor due to induction.
- Changing the DC excitation does not affect the speed of the motor.
- Selsyn motors are used to provide position control and angular feedback information.
- Although selsyn motors contain three-phase windings, they operate on single-phase AC.
- A differential selsyn unit can be used to determine the algebraic sum of the rotation of two other selsyn units.

Review Questions

1. What are the three basic types of three-phase motors?
2. What is the principle of operation of all three-phase motors?
3. What is synchronous speed?
4. What two factors determine synchronous speed?

5. Name three factors that cause the magnetic field to rotate.
6. Name three factors that determine the torque produced by an induction motor.
7. Is the synchronous motor an induction motor?
8. What is the amortisseur winding?
9. Why must a synchronous motor never be started when DC excitation is applied to the rotor?
10. Name three characteristics that make the synchronous motor different from an induction motor.
11. What is the function of the field-discharge resistor?
12. Why can an induction motor never operate at synchronous speed?
13. A squirrel-cage induction motor is operating at 1175 rpm and producing a torque of 22 lb-ft. What is the horsepower output of the motor?
14. A wattmeter measures the input power of the motor in Question 13 to be 5650 W. What is the efficiency of the motor?
15. What is the difference between a squirrel-cage motor and a wound-rotor motor?
16. What is the advantage of the wound-rotor motor over the squirrel-cage motor?
17. Name three factors that determine the amount of voltage induced in the rotor of a wound-rotor motor.
18. Why will the rotor of a wound-rotor motor not turn if the rotor circuit is left open with no resistance connected to it?
19. Why is the starting torque per A of starting current of a wound-rotor motor higher than that of a squirrel-cage motor although the starting current is less?
20. When is a synchronous motor a synchronous condenser?
21. What determines when a synchronous motor is at normal excitation?
22. How can a synchronous motor be made to have a leading power factor?
23. Is the excitation current of a synchronous motor AC or DC?
24. How is the speed of a consequent-pole squirrel-cage motor changed?
25. A three-phase squirrel-cage motor is connected to a 60-Hz line. The full-load speed is 870 rpm. How many poles per phase does the stator have?

SECTION XIV AC Machines

Practical Applications

You are working as a plant electrician. It is your job to install a 300-hp three-phase squirrel-cage induction motor. The supply voltage is 480 V. The power company has determined that the maximum amount of starting current that can be permitted by any motor in the plant is 3000 A. The motor nameplate is as follows:

Phase: 3	FLA: 352
Volts: 480	RPM: 1755
Frame: XXX	Code: L

Will it be possible to start this motor across the line, or will it be necessary to use a reduced-voltage starter to reduce starting current? ■

Practical Applications

You have been given the task of connecting a nine-lead three-phase dual-voltage motor to a 240-V line. You discover that the nameplate has been painted and you cannot see the connection diagram. To make the proper connections, you must know if the motor stator winding is connected wye or delta. How could you determine this using an ohmmeter? Explain your answer. ■

Practical Applications

You are an electrician working in an industrial plant. A 30-hp three-phase squirrel-cage motor keeps tripping out on overload. The motor is connected to a 240-volt line. Voltage measurements indicate the following voltages between the different phases: A–B = 276 volts, B–C = 221 volts, and A–C = 267 volts. Determine the percentage of heat rise in the winding with the highest current. ■

Unit 31
Single-Phase Motors

OUTLINE

31–1	Single-Phase Motors
31–2	Split-Phase Motors
31–3	Resistance-Start Induction-Run Motors
31–4	Capacitor-Start Induction-Run Motors
31–5	Dual-Voltage Split-Phase Motors
31–6	Determining the Direction of Rotation for Split-Phase Motors
31–7	Capacitor-Start Capacitor-Run Motors
31–8	Shaded-Pole Induction Motors
31–9	Multispeed Motors
31–10	Repulsion-Type Motors
31–11	Construction of Repulsion Motors
31–12	Repulsion-Start Induction-Run Motors
31–13	Repulsion-Induction Motors
31–14	Single-Phase Synchronous Motors
31–15	Stepping Motors
31–16	Universal Motors

KEY TERMS

Centrifugal switch
Compensating winding
Conductive compensation
Consequent-pole motor
Holtz motor
Inductive compensation
Multispeed motors
Neutral plane
Repulsion motor
Run winding
Shaded-pole induction motor
Shading coil
Split-phase motors
Start winding
Stepping motors
Synchronous motors
Two-phase
Universal motor
Warren motor

Why You Need to Know

Single-phase motors are used almost exclusively in residential applications and to operate loads that require fractional horsepower motors in industrial and commercial locations. Many of these motors you will recognize from everyday life and may have wondered how they work. Unlike three-phase motors, there are many different types of single-phase motors and they do not all operate on the same principle. There are some that operate on the principle of a rotating magnetic field, but others do not. Some single-phase motors are designed to operate at more than one speed. This unit

- presents several different types of single-phase motors and explains how they operate.
- explains how to determine the appropriate motor to be used under a given situation by evaluating the operating principles of each.

Objectives

After studying this unit, you should be able to

- list the different types of split-phase motors.
- discuss the operation of split-phase motors.
- reverse the direction of rotation of a split-phase motor.
- discuss the operation of multispeed split-phase motors.
- discuss the operation of shaded-pole-type motors.
- discuss the operation of repulsion-type motors.
- discuss the operation of stepping motors.
- discuss the operation of universal motors.

Preview

Although most of the large motors used in industry are three phase, at times single-phase motors must be used. Single-phase motors are used almost exclusively to operate home appliances such as air conditioners, refrigerators, well pumps, and fans. They are generally designed to operate on 120 volts or 240 volts. They range in size from fractional horsepower to several horsepower, depending on the application. ■

31-1 Single-Phase Motors

In Unit 30, it was stated that there are three basic types of three-phase motors and that all operate on the principle of a rotating magnetic field. Although that is true for three-phase motors, it is not true for single-phase motors. There are not only many different types of single-phase motors, but they also have different operating principles.

31-2 Split-Phase Motors

Split-phase motors fall into three general classifications:

1. The resistance-start induction-run motor
2. The capacitor-start induction-run motor
3. The capacitor-start capacitor-run motor

Although all these motors have different operating characteristics, they are similar in construction and use the same operating principle. Split-phase motors

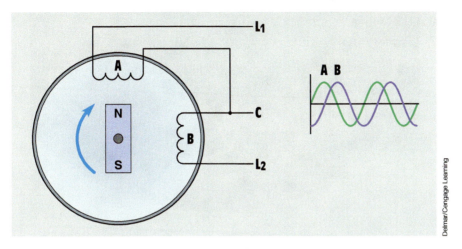

FIGURE 31–1 A two-phase alternator produces voltages that are 90° out of phase with each other.

receive their name from the manner in which they operate. Like three-phase motors, split-phase motors operate on the principle of a rotating magnetic field. A rotating magnetic field, however, cannot be produced with only one phase. Split-phase motors therefore split the current flow through two separate windings to simulate a two-phase power system. A rotating magnetic field can be produced with a two-phase system.

The Two-Phase System

In some parts of the world, **two-phase** power is produced. A two-phase system is produced by having an alternator with two sets of coils wound 90° apart *(Figure 31–1)*. The voltages of a two-phase system are therefore 90° out of phase with each other. These two out-of-phase voltages can be used to produce a rotating magnetic field in a manner similar to that of producing a rotating magnetic field with the voltages of a three-phase system. Because there have to be two voltages or currents out of phase with each other to produce a rotating magnetic field, split-phase motors use two separate windings to create a phase difference between the currents in each of these windings. These motors literally split one phase and produce a second phase, hence the name split-phase motor.

Stator Windings

The stator of a split-phase motor contains two separate windings, the **start winding** and the **run winding.** The start winding is made of small wire and is placed near the top of the stator core. The run winding is made of relatively large wire and is placed in the bottom of the stator core. *Figures 31–2A and B*

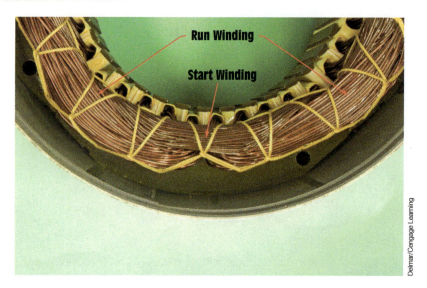

FIGURE 31–2A Stator winding of a resistance-start induction-run motor. The start winding contains much smaller wire than the run winding.

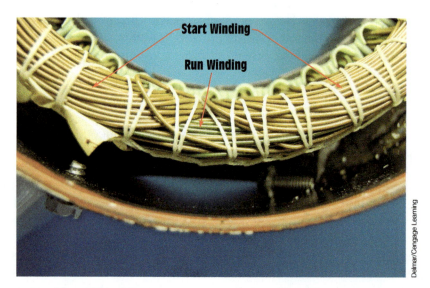

FIGURE 31–2B Stator winding of a capacitor-start capacitor-run motor. The wire size is the same for both start and run windings.

are photographs of two split-phase stators. The stator in *Figure 31–2A* is used for a resistance-start induction-run motor or a capacitor-start induction-run motor. The stator in *Figure 31–2B* is used for a capacitor-start capacitor-run motor. Both stators contain four poles, and the start winding is placed at a 90° angle from the run winding.

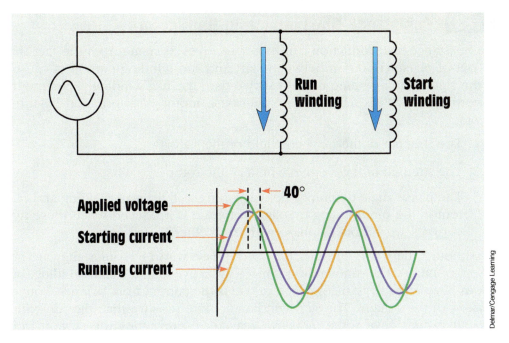

FIGURE 31–3 The start and run windings are connected in parallel with each other.

Notice the difference in size and position of the two windings of the stator shown in *Figure 31–2A*. The start winding is made from small wire and placed near the top of the stator core. This causes it to have a higher resistance than the run winding. The start winding is located between the poles of the run winding. The run winding is made with larger wire and placed near the bottom of the core. This gives it higher inductive reactance and less resistance than the start winding. These two windings are connected in parallel with each other *(Figure 31–3)*.

When power is applied to the stator, current flows through both windings. Because the start winding is more resistive, the current flow through it is more in phase with the applied voltage than the current flow through the run winding. The current flow through the run winding lags the applied voltage due to inductive reactance. These two out-of-phase currents are used to create a rotating magnetic field in the stator. The speed of this rotating magnetic field is called synchronous speed and is determined by the same two factors that determined the synchronous speed for a three-phase motor:

1. Number of stator poles per phase
2. Frequency of the applied voltage

31-3 Resistance-Start Induction-Run Motors

The resistance-start induction-run motor receives its name from the fact that the out-of-phase condition between start and run winding current is caused by the start winding being more resistive than the run winding. The amount of starting torque produced by a split-phase motor is determined by three factors:

1. The strength of the magnetic field of the stator

2. The strength of the magnetic field of the rotor

3. The phase angle difference between current in the start winding and current in the run winding (Maximum torque is produced when these two currents are 90° out of phase with each other.)

Although these two currents are out of phase with each other, they are not 90° out of phase. The run winding is more inductive than the start winding, but it does have some resistance, which prevents the current from being 90° out of phase with the voltage. The start winding is more resistive than the run winding, but it does have some inductive reactance, preventing the current from being in phase with the applied voltage. Therefore, a phase angle difference of 35° to 40° is produced between these two currents, resulting in a rather poor starting torque *(Figure 31–4)*.

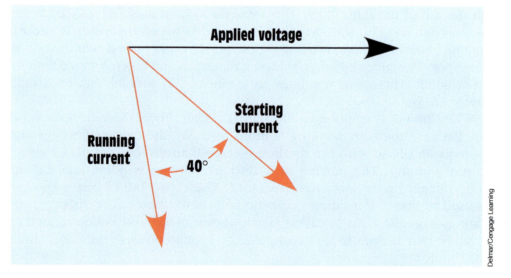

FIGURE 31-4 Running current and starting current are 35° to 40° out of phase with each other.

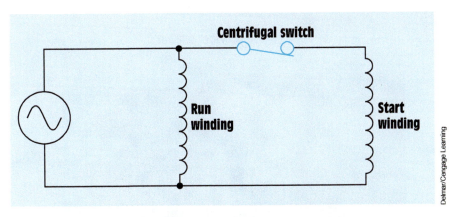

FIGURE 31-5 A centrifugal switch is used to disconnect the start winding from the circuit.

Disconnecting the Start Winding

A stator rotating magnetic field is necessary only to start the rotor turning. Once the rotor has accelerated to approximately 75% of rated speed, the start winding can be disconnected from the circuit and the motor will continue to operate with only the run winding energized. Motors that are not hermetically sealed (most refrigeration and air-conditioning compressors are hermetically sealed) use a **centrifugal switch** to disconnect the start windings from the circuit. The contacts of the centrifugal switch are connected in series with the start winding *(Figure 31–5)*. The centrifugal switch contains a set of spring-loaded weights. When the shaft is not turning, the springs hold a fiber washer in contact with the movable contact of the switch *(Figure 31–6)*. The fiber washer causes the movable contact to complete a circuit with a stationary contact.

When the rotor accelerates to about 75% of rated speed, centrifugal force causes the weights to overcome the force of the springs. The fiber washer retracts and permits the contacts to open and disconnect the start winding from the circuit *(Figure 31–7)*. The start winding of this type motor is intended to be energized only during the period of time that the motor is actually starting. If the start winding is not disconnected, it will be damaged by excessive current flow.

Starting Relays

Resistance-start induction-run and capacitor-start induction-run motors are sometimes hermetically sealed, such as with air-conditioning and refrigeration compressors. When these motors are hermetically sealed, a centrifugal switch cannot be used to disconnect the start winding. Some device that can be mounted externally must be used to disconnect the start windings from

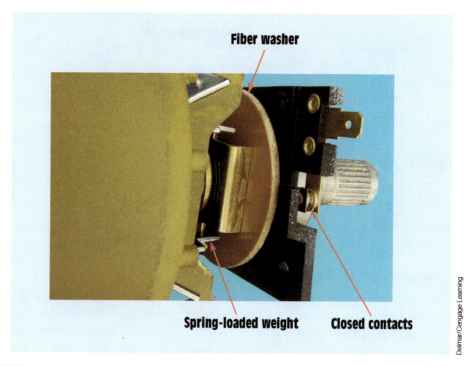

FIGURE 31–6 The centrifugal switch is closed when the rotor is not turning.

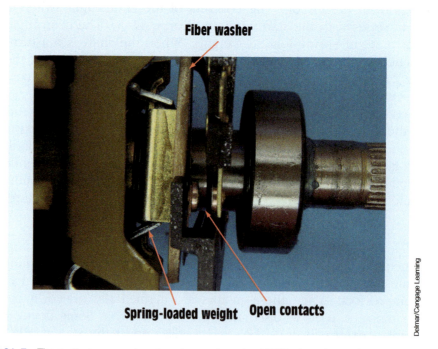

FIGURE 31–7 The contact opens when the rotor reaches about 75% of rated speed.

the circuit. Starting relays are used to perform this function. There are three basic types of starting relays used with the resistance-start and capacitor-start motors:

1. Hot-wire relay
2. Current relay
3. Solid-state starting relay

The *hot-wire relay* functions as both a starting relay and an overload relay. In the circuit shown in *Figure 31–8*, it is assumed that a thermostat controls the operation of the motor. When the thermostat closes, current flows through a resistive wire and two normally closed contacts connected to the start and run windings of the motor. The high starting current of the motor rapidly heats the resistive wire, causing it to expand. The expansion of the wire causes the spring-loaded start winding contact to open and disconnect the start winding from the circuit, reducing motor current. If the motor is not overloaded, the resistive wire never becomes hot enough to cause the overload contact to open and the motor continues to run. If the motor should become overloaded, however, the resistive wire expands enough to open the overload contact and

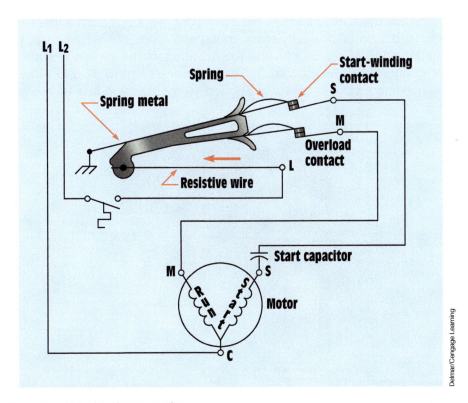

FIGURE 31–8 Hot-wire relay connection.

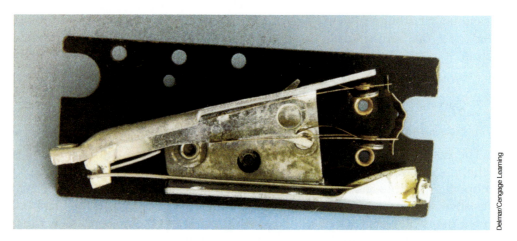

FIGURE 31–9 Hot-wire type of starting relay.

disconnect the motor from the line. A photograph of a hot-wire starting relay is shown in *Figure 31–9*.

The *current relay* also operates by sensing the amount of current flow in the circuit. This type of relay operates on the principle of a magnetic field instead of expanding metal. The current relay contains a coil with a few turns of large wire and a set of normally open contacts *(Figure 31–10)*. The coil of the relay is connected in series with the run winding of the motor, and the contacts are connected in series with the start winding *(Figure 31–11)*. When the thermostat contact closes, power is applied to the run winding of the motor.

FIGURE 31–10 Current type of starting relay.

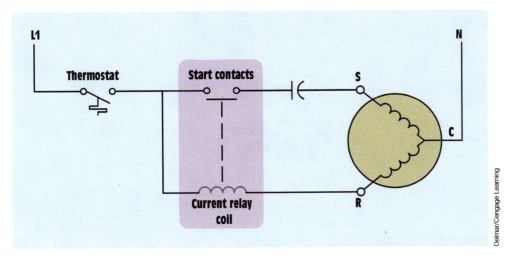

FIGURE 31–11 Current relay connection.

Because the start winding is open, the motor cannot start, causing a high current to flow in the run winding circuit. This high current flow produces a strong magnetic field in the coil of the relay, causing the normally open contacts to close and connect the start winding to the circuit. When the motor starts, the run-winding current is greatly reduced, permitting the start contacts to reopen and disconnect the start winding from the circuit.

The *solid-state starting relay (Figure 31–12)* performs the same basic function as the current relay and in many cases is replacing both the current relay and the centrifugal switch. The solid-state starting relay is generally more reliable and less expensive than the current relay or the centrifugal switch. The solid-state

FIGURE 31–12 Solid-state starting relay.

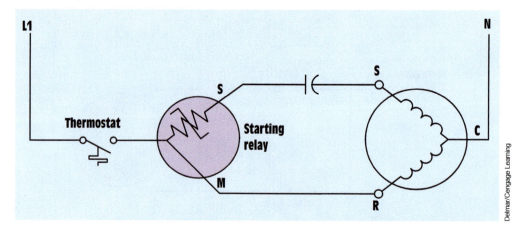

FIGURE 31–13 Solid-state starting relay connection.

starting relay is actually an electronic component known as a *thermistor*. A thermistor is a device that exhibits a change of resistance with a change of temperature. This particular thermistor has a positive coefficient of temperature, which means that when its temperature increases, its resistance increases also. The schematic diagram in *Figure 31–13* illustrates the connection of the solid-state starting relay. The thermistor is connected in series with the start winding of the motor. When the motor is not in operation, the thermistor is at a low temperature and its resistance is low, typically 3 or 4 ohms. When the thermostat contact closes, current flows to both the run and start windings of the motor. The current flowing through the thermistor causes an increase in temperature. This increased temperature causes the resistance of the thermistor to suddenly change to a high value of several thousand ohms. The change of temperature is so sudden that it has the effect of opening a set of contacts. Although the start winding is never completely disconnected from the powerline, the amount of current flow though it is very small, typically 0.03 to 0.05 amperes, and does not affect the operation of the motor. This small amount of *leakage current* maintains the temperature of the thermistor and prevents it from returning to a low value of resistance. After the motor is disconnected from the powerline, a cooldown time of two to three minutes should be allowed to permit the thermistor to return to a low resistance before the motor is restarted.

Relationship of Stator and Rotor Fields

The split-phase motor contains a squirrel-cage rotor very similar to those used with three-phase squirrel-cage motors *(Figure 31–14)*. When power is connected to the stator windings, the rotating magnetic field induces a voltage into the bars of the squirrel-cage rotor. The induced voltage causes current to flow in the rotor, and a magnetic field is produced around the rotor bars. The magnetic field of the rotor is attracted to the stator field, and the rotor begins to

FIGURE 31-14 Squirrel-cage rotor used in a split-phase motor.

turn in the direction of the rotating magnetic field. After the centrifugal switch opens, only the run winding induces voltage into the rotor. This induced voltage is in phase with the stator current. The inductive reactance of the rotor is high, causing the rotor current to be almost 90° out of phase with the induced voltage. This causes the pulsating magnetic field of the rotor to lag the pulsating magnetic field of the stator by 90°. Magnetic poles, located midway between the stator poles, are created in the rotor *(Figure 31–15)*. These two pulsating magnetic fields produce a rotating magnetic field of their own, and the rotor continues to rotate.

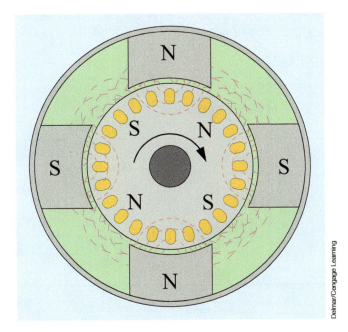

FIGURE 31-15 A rotating magnetic field is produced by the stator and rotor flux.

Direction of Rotation

The direction of rotation for the motor is determined by the direction of rotation of the rotating magnetic field created by the run and start windings when the motor is first started. The direction of motor rotation can be changed by reversing the connection of either the start winding or the run winding, but not both. If the start winding is disconnected, the motor can be operated in either direction by manually turning the rotor shaft in the desired direction of rotation.

31–4 Capacitor-Start Induction-Run Motors

The capacitor-start induction-run motor is very similar in construction and operation to the resistance-start induction-run motor. The capacitor-start induction-run motor, however, has an AC electrolytic capacitor connected in series with the centrifugal switch and start winding *(Figure 31–16)*. Although the running characteristics of the capacitor-start induction-run motor and the resistance-start induction-run motor are identical, the starting characteristics are not. The capacitor-start induction-run motor produces a starting torque that is substantially higher than that of the resistance-start induction-run motor. Recall that one of the factors that determines the starting torque for a split-phase motor is the phase angle difference between start-winding current and run-winding current. The starting torque of a resistance-start induction-run motor is low because the phase angle difference between these two currents is only about 40° *(Figure 31–4)*.

When a capacitor of the proper size is connected in series with the start winding, it causes the start-winding current to lead the applied voltage. This

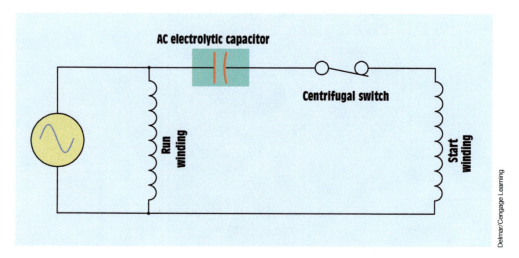

FIGURE 31–16 An AC electrolytic capacitor is connected in series with the start winding.

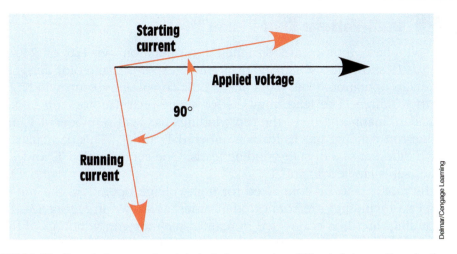

FIGURE 31–17 Run-winding current and start-winding current are 90° out of phase with each other.

leading current produces a 90° phase shift between run-winding current and start-winding current *(Figure 31–17)*. Maximum starting torque is developed at this point.

Although the capacitor-start induction-run motor has a high starting torque, the motor should not be started more than about eight times per hour. Frequent starting can damage the start capacitor due to overheating. If the capacitor must be replaced, care should be taken to use a capacitor of the correct microfarad rating. If a capacitor with too little capacitance is used, the starting current will be less than 90° out of phase with the running current, and the starting torque will be reduced. If the capacitance value is too great, the starting current will be more than 90° out of phase with the running current, and the starting torque will again be reduced. A capacitor-start induction-run motor is shown in *Figure 31–18*.

FIGURE 31–18 Capacitor-start induction-run motor.

31–5 Dual-Voltage Split-Phase Motors

Many split-phase motors are designed for operation on 120 or 240 volts. *Figure 31–19* shows the schematic diagram of a split-phase motor designed for dual-voltage operation. This particular motor contains two run windings and two start windings. The lead numbers for single-phase motors are numbered in a standard manner. One of the run windings has lead numbers of T_1 and T_2. The other run winding has its leads numbered T_3 and T_4. This particular motor uses two different sets of start-winding leads. One set is labeled T_5 and T_6, and the other set is labeled T_7 and T_8.

If the motor is to be connected for high-voltage operation, the run windings and start windings are connected in series, as shown in *Figure 31–20*. The start windings are then connected in parallel with the run windings. If the opposite direction of rotation is desired, T_5 and T_8 are changed.

For low-voltage operation, the windings must be connected in parallel, as shown in *Figure 31–21*. This connection is made by first connecting the run windings in parallel by hooking T_1 and T_3 together and T_2 and T_4 together. The start windings are paralleled by connecting T_5 and T_7 together and T_6 and T_8 together. The start windings are then connected in parallel with the run windings. If the opposite direction of rotation is desired, T_5 and T_6 should be reversed along with T_7 and T_8.

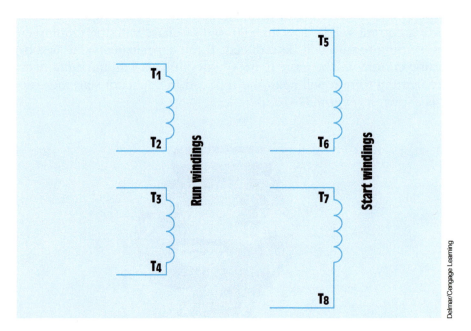

FIGURE 31–19 Dual-voltage windings for a split-phase motor.

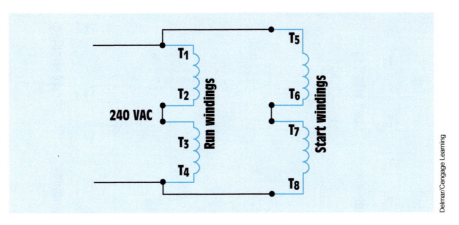

FIGURE 31–20 High-voltage connection for a split-phase motor with two run and two start windings.

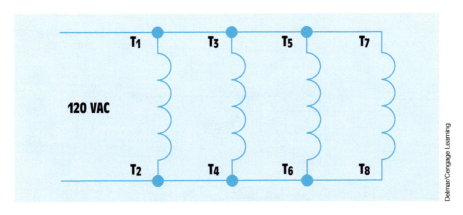

FIGURE 31–21 Low-voltage connection for a split-phase motor with two run and two start windings.

Not all dual-voltage single-phase motors contain two sets of start windings. *Figure 31–22* shows the schematic diagram of a motor that contains two sets of run windings and only one start winding. In this illustration, the start winding is labeled T_5 and T_6. Some motors, however, identify the start winding by labeling it T_5 and T_8, as shown in *Figure 31–23*.

Regardless of which method is used to label the terminal leads of the start winding, the connection is the same. If the motor is to be connected for high-voltage operation, the run windings are connected in series and the start winding is connected in parallel with one of the run windings (*Figure 31–24*). In this type of motor, each winding is rated at 120 volts. If the run windings are connected in series across 240 volts, each winding has a voltage drop of 120 volts. By connecting the start winding in parallel across only one run winding, it

FIGURE 31–22 Dual-voltage motor with one start winding labeled T_5 and T_6.

FIGURE 31–23 Dual-voltage motor with one start winding labeled T_5 and T_8.

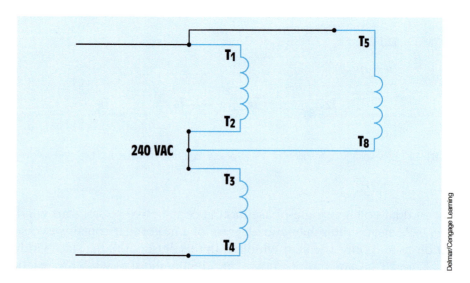

FIGURE 31–24 High-voltage connection with one start winding.

receives only 120 volts when power is applied to the motor. If the opposite direction of rotation is desired, T_5 and T_8 should be changed.

If the motor is to be operated on low voltage, the windings are connected in parallel as shown in *(Figure 31–25)*. Because all windings are connected in parallel, each receives 120 volts when power is applied to the motor.

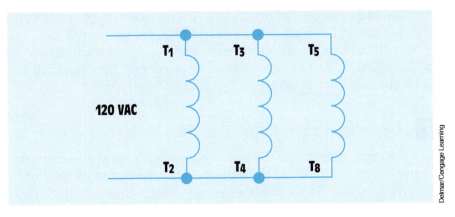

FIGURE 31–25 Low-voltage connection for a split-phase motor with one start winding.

31–6 Determining the Direction of Rotation for Split-Phase Motors

The direction of rotation of a single-phase motor can generally be determined when the motor is connected. The direction of rotation is determined by facing the back or rear of the motor. *Figure 31–26* shows a connection diagram for rotation. If clockwise rotation is desired, T_5 should be connected to T_1. If counterclockwise rotation is desired, T_8 (or T_6) should be connected to T_1. This connection diagram assumes that the motor contains two sets of run and two sets of start windings. The type of motor used determines the actual connection. For example, *Figure 31–24* shows the connection of a motor with two

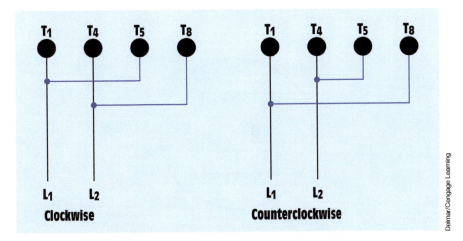

FIGURE 31–26 Determining direction of rotation for a split-phase motor.

run windings and only one start winding. If this motor were to be connected for clockwise rotation, terminal T_5 would have to be connected to T_1, and terminal T_8 would have to be connected to T_2 and T_3. If counterclockwise rotation is desired, terminal T_8 would have to be connected to T_1, and terminal T_5 would have to be connected to T_2 and T_3.

31–7 Capacitor-Start Capacitor-Run Motors

Although the capacitor-start capacitor-run motor is a split-phase motor, it operates on a different principle than the resistance-start induction-run motor or the capacitor-start induction-run motor. The capacitor-start capacitor-run motor is designed in such a manner that its start winding remains energized at all times. A capacitor is connected in series with the winding to provide a continuous leading current in the start winding *(Figure 31–27)*. Because the start winding remains energized at all times, no centrifugal switch is needed to disconnect the start winding as the motor approaches full speed. The capacitor used in this type of motor is generally of the oil-filled type because it is intended for continuous use. An exception to this general rule is small fractional-horsepower motors used in reversible ceiling fans. These fans have a low current draw and use an AC electrolytic capacitor to help save space.

The capacitor-start capacitor-run motor actually operates on the principle of a rotating magnetic field in the stator. Because both run and start windings remain energized at all times, the stator magnetic field continues to rotate and the motor operates as a two-phase motor. This motor has excellent starting and running torque. It is quiet in operation and has a high efficiency. Because the capacitor remains connected in the circuit at all times, the motor power factor is close to unity.

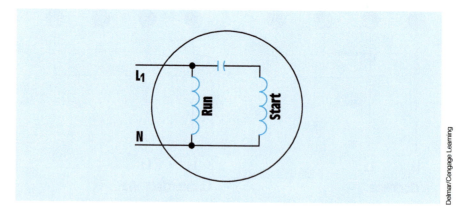

FIGURE 31–27 A capacitor-start capacitor-run motor.

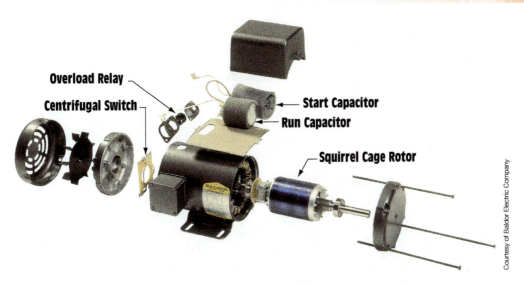

FIGURE 31–28 Capacitor-start capacitor-run motor with additional starting capacitor.

Although the capacitor-start capacitor-run motor does not require a centrifugal switch to disconnect the capacitor from the start winding, some motors use a second capacitor during the starting period to help improve starting torque *(Figure 31–28)*. A good example of this can be found on the compressor of a central air-conditioning unit designed for operation on single-phase power. If the motor is not hermetically sealed, a centrifugal switch will be used to disconnect the start capacitor from the circuit when the motor reaches approximately 75% of rated speed. Hermetically sealed motors, however, must use some type of external switch to disconnect the start capacitor from the circuit.

The capacitor-start capacitor-run motor, or permanent split-capacitor motor as it is generally referred to in the air-conditioning and refrigeration industry, generally employs a potential starting relay to disconnect the starting capacitor when a centrifugal switch cannot be used. The potential starting relay *(Figure 31–29A and B)* operates by sensing an increase in the voltage developed in the start winding when the motor is operating. A schematic diagram of a potential starting relay circuit is shown in *Figure 31–30*. In this circuit, the potential relay is used to disconnect the starting capacitor from the circuit when the motor reaches about 75% of its full speed. The starting-relay coil, SR, is connected in parallel with the start winding of the motor. A normally closed SR contact is connected in series with the starting capacitor. When the thermostat contact closes, power is applied to both the run and start windings. At this point in time, both the start and run capacitors are connected in the circuit.

As the rotor begins to turn, its magnetic field induces a voltage into the start winding, producing a higher voltage across the start winding than the applied voltage. When the motor has accelerated to about 75% of its full

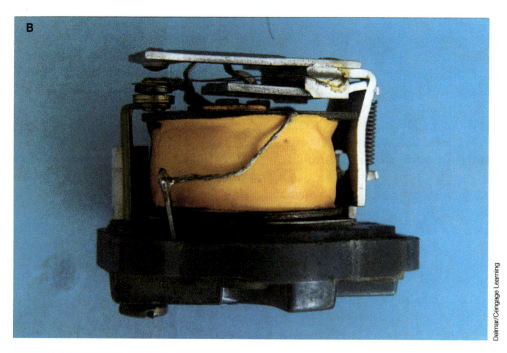

FIGURE 31–29A AND B Potential starting relays.

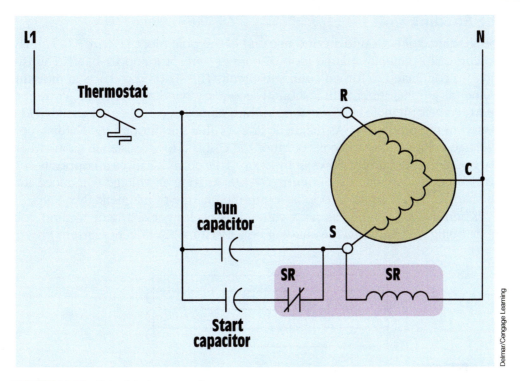

FIGURE 31-30 Potential relay connection.

speed, the voltage across the start winding is high enough to energize the coil of the potential relay. This causes the normally closed SR contact to open and disconnect the start capacitor from the circuit. Because the start winding of this motor is never disconnected from the powerline, the coil of the potential starting relay remains energized as long as the motor is in operation.

31-8 Shaded-Pole Induction Motors

The **shaded-pole induction motor** is popular because of its simplicity and long life. This motor contains no start windings or centrifugal switch. It contains a squirrel-cage rotor and operates on the principle of a rotating magnetic field. The rotating magnetic field is created by a **shading coil** wound on one side of each pole piece. Shaded-pole motors are generally fractional-horsepower motors and are used for low-torque applications such as operating fans and blowers.

The Shading Coil

The shading coil is wound around one end of the pole piece *(Figure 31–31)*. The shading coil is actually a large loop of copper wire or a copper band. The two ends are connected to form a complete circuit. The shading coil acts in the same manner as a transformer with a shorted secondary winding. When the current of the AC waveform increases from zero toward its positive peak, a magnetic field is created in the pole piece. As magnetic lines of flux cut through the shading coil, a voltage is induced in the coil. Because the coil is a low-resistance short circuit, a large amount of current flows in the loop. This current causes an opposition to the change of magnetic flux *(Figure 31–32)*. As long as voltage is induced into the shading coil, there is an opposition to the change of magnetic flux.

When the AC reaches its peak value, it is no longer changing and no voltage is being induced into the shading coil. Because there is no current flow in

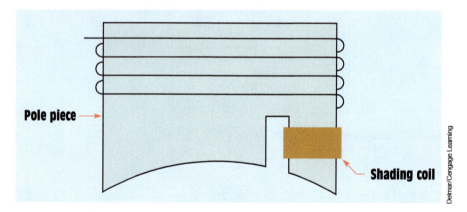

FIGURE 31–31 A shaded pole.

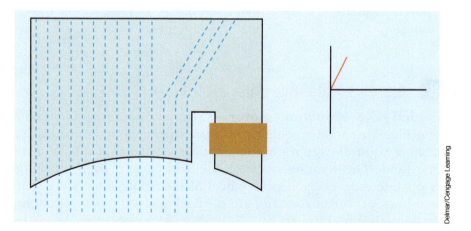

FIGURE 31–32 The shading coil opposes a change of flux as current increases.

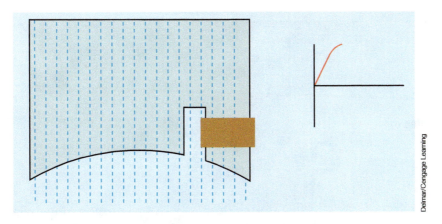

FIGURE 31–33 There is opposition to magnetic flux when the current is not changing.

the shading coil, there is no opposition to the magnetic flux. The magnetic flux of the pole piece is now uniform across the pole face *(Figure 31–33)*.

When the AC begins to decrease from its peak value back toward zero, the magnetic field of the pole piece begins to collapse. A voltage is again induced into the shading coil. This induced voltage creates a current that opposes the change of magnetic flux *(Figure 31–34)*. This causes the magnetic flux to be concentrated in the shaded section of the pole piece.

When the AC passes through zero and begins to increase in the negative direction, the same set of events happens except that the polarity of the magnetic field is reversed. If these events were to be viewed in rapid order, the magnetic field would be seen to rotate across the face of the pole piece. A pole piece with a shading coil is shown in *Figure 31–35*.

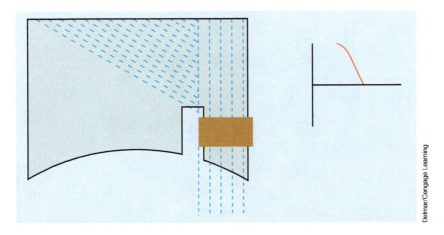

FIGURE 31–34 The shading coil opposes a change of flux when the current decreases.

FIGURE 31–35   The shading coil is a large copper conductor wound around one side of the pole piece.

Speed

The speed of the shaded-pole induction motor is determined by the same factors that determine the synchronous speed of other induction motors: frequency and number of stator poles. Shaded-pole motors are commonly wound as four- or six-pole motors. *Figure 31–36* shows a drawing of a four-pole shaded-pole induction motor.

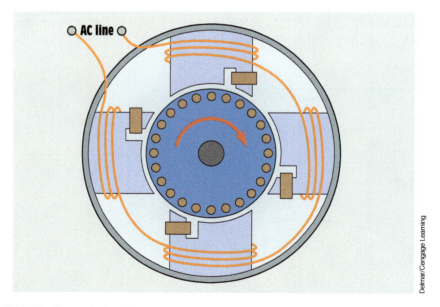

FIGURE 31–36   Four-pole shaded-pole induction motor.

FIGURE 31–37 Stator winding and rotor of a shaded-pole induction motor.

General Operating Characteristics

The shaded-pole motor contains a standard squirrel-cage rotor. The amount of torque produced is determined by the strength of the magnetic field of the stator, the strength of the magnetic field of the rotor, and the phase angle difference between rotor and stator flux. The shaded-pole induction motor has low starting and running torque.

The direction of rotation is determined by the direction in which the rotating magnetic field moves across the pole face. The rotor turns in the direction shown by the arrow in *Figure 31–36*. The direction can be changed by removing the stator winding and turning it around. This is not a common practice, however. As a general rule, the shaded-pole induction motor is considered to be nonreversible. *Figure 31–37* shows a photograph of the stator winding and rotor of a shaded-pole induction motor.

31–9 Multispeed Motors

There are two basic types of single-phase **multispeed motors.** One is the consequent-pole type and the other is a specially wound *capacitor-start* capacitor-run motor or shaded-pole induction motor. The single-phase **consequent-pole motor** operates in the same basic way as the three-phase consequent pole discussed in Unit 30. The speed is changed by reversing the current flow through alternate poles and increasing or decreasing the total number of stator poles. The consequent-pole motor is used where high running torque must be maintained at different speeds. A good example of where this type of motor is used is in two-speed compressors for central air-conditioning units.

Multispeed Fan Motors

Multispeed fan motors have been used for many years. These motors are generally wound for two to five steps of speed and operate fans and squirrel-cage blowers. A schematic drawing of a three-speed motor is shown in *Figure 31–38*. Notice that the run winding has been tapped to produce low, medium, and high speed. The start winding is connected in parallel with the run-winding section. The other end of the start-winding lead is connected to an external oil-filled capacitor. This motor obtains a change of speed by inserting inductance in series with the run winding. The actual run winding for this motor is between the terminals marked *high* and *common*. The winding shown between *high* and *medium* is connected in series with the main run winding. When the rotary switch is connected to the medium speed position, the inductive reactance of this coil limits the amount of current flow through the run winding. When the current of the run winding is reduced, the strength of the magnetic field of the run winding is reduced and the motor produces less torque. This causes a greater amount of slip and the motor speed to decrease.

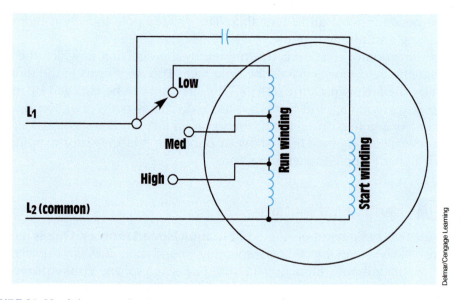

FIGURE 31–38   A three-speed motor.

If the rotary switch is changed to the *low* position, more inductance is inserted in series with the run winding. This causes less current to flow through the run winding and another reduction in torque. When the torque is reduced, the motor speed decreases again.

Common speeds for a four-pole motor of this type are 1625, 1500, and 1350 rpm. Notice that this motor does not have wide ranges between speeds as would be the case with a consequent-pole motor. Most induction motors would overheat and damage the motor winding if the speed were reduced to this extent. This type of motor, however, has much higher impedance windings than most other motors. The run windings of most split-phase motors have a wire resistance of 1 to 4 ohms. This motor generally has a resistance of 10 to 15 ohms in its run winding. It is the high impedance of the windings that permits the motor to be operated in this manner without damage.

Because this motor is designed to slow down when load is added, it is not used to operate high-torque loads. This type of motor is generally used to operate only low-torque loads such as fans and blowers.

31–10 Repulsion-Type Motors

There are three basic repulsion-type motors:

1. The repulsion motor
2. The repulsion-start induction-run motor
3. The repulsion-induction motor

Each of these three types has different operating characteristics.

31–11 Construction of Repulsion Motors

A **repulsion motor** operates on the principle that like magnetic poles repel each other, not on the principle of a rotating magnetic field. The stator of a repulsion motor contains only a run winding very similar to that used in the split-phase motor. Start windings are not necessary. The rotor is actually called an armature because it contains a slotted metal core with windings placed in the slots. The windings are connected to a commutator. A set of brushes makes contact with the surface of the commutator bars. The entire assembly looks very much like a DC armature and brush assembly. One difference, however, is that the brushes of the repulsion motor are shorted together. Their function is to provide a current path through certain parts of the armature, not to provide power to the armature from an external source.

Operation

Although the repulsion motor does not operate on the principle of a rotating magnetic field, it is an induction motor. When AC power is connected to the stator winding, a magnetic field with alternating polarities is produced in the poles. This alternating field induces a voltage into the windings of the armature. When the brushes are placed in the proper position, current flows through the armature windings, producing a magnetic field of the same polarity in the armature. The armature magnetic field is repelled by the stator magnetic field, causing the armature to rotate. Repulsion motors contain the same number of brushes as there are stator poles. Repulsion motors are commonly wound for four, six, or eight poles.

Brush Position

The position of the brushes is very important. Maximum torque is developed when the brushes are placed 15° on either side of the pole pieces. *Figure 31–39* shows the effect of having the brushes placed at a 90° angle to the pole pieces. When the brushes are in this position, a circuit is completed between the coils located at a right angle to the poles. In this position, there is no induced voltage in the armature windings and no torque is produced by the motor.

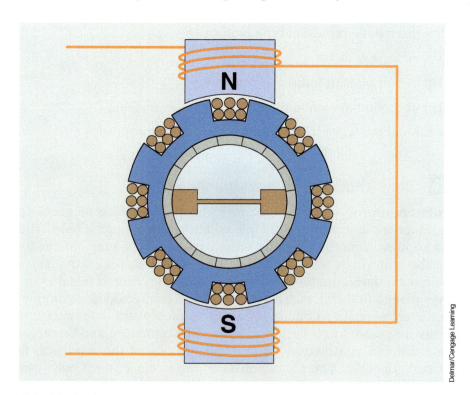

FIGURE 31–39 Brushes are placed at a 90° angle to the poles.

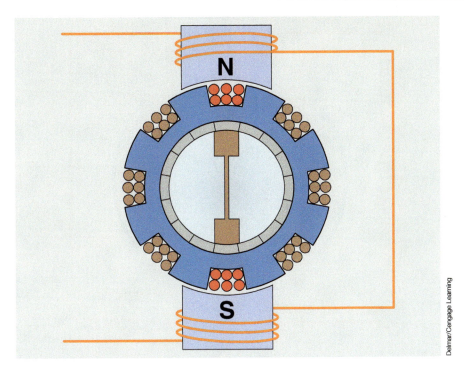

FIGURE 31–40 The brushes are set at a 0° angle to the pole pieces.

In *Figure 31–40*, the brushes have been moved to a position so that they are in line with the pole pieces. In this position, a large amount of current flows through the coils directly under the pole pieces. This current produces a magnetic field of the same polarity as the pole piece. Because the magnetic field produced in the armature is at a 0° angle to the magnetic field of the pole piece, no twisting or turning force is developed and the armature does not turn.

In *Figure 31–41*, the brushes have been shifted in a clockwise direction so that they are located 15° from the pole piece. The induced voltage in the armature winding produces a magnetic field of the same polarity as the pole piece. The magnetic field of the armature is repelled by the magnetic field of the pole piece, and the armature turns in the clockwise direction.

In *Figure 31–42*, the brushes have been shifted counterclockwise to a position 15° from the center of the pole piece. The magnetic field developed in the armature again repels the magnetic field of the pole piece, and the armature turns in the counterclockwise direction.

The direction of armature rotation is determined by the setting of the brushes. The direction of rotation for any type of repulsion motor is changed by setting the brushes 15° on either side of the pole pieces. Repulsion-type motors have the highest starting torque of any single-phase motor. The speed of a repulsion motor, not to be confused with the repulsion-start induction-run

962 SECTION XIV AC Machines

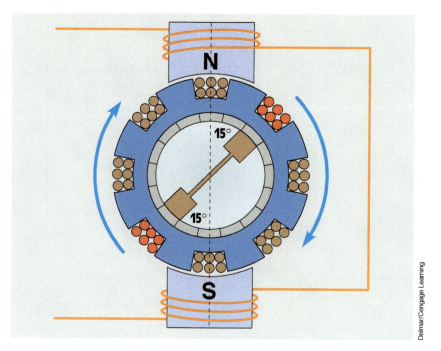

FIGURE 31–41 The brushes have been shifted clockwise 15°.

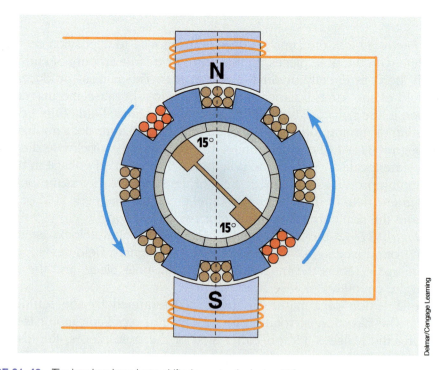

FIGURE 31–42 The brushes have been shifted counterclockwise 15°.

motor or the repulsion-induction motor, can be varied by changing the AC voltage supplying power for the motor. The repulsion motor has excellent starting and running torque but can exhibit unstable speed characteristics. The repulsion motor can race to very high speed if operated with no mechanical load connected to the shaft.

31–12 Repulsion-Start Induction-Run Motors

The repulsion-start induction-run motor starts as a repulsion motor but runs like a squirrel-cage motor. There are two types of repulsion-start induction-run motors:

1. The brush-riding type
2. The brush-lifting type

The brush-riding type uses an axial commutator *(Figure 31–43)*. The brushes ride against the commutator segments at all times when the motor is in operation. After the motor has accelerated to approximately 75% of its full-load speed, centrifugal force causes copper segments of a short-circuiting ring to overcome the force of a spring *(Figure 31–44)*. The segments sling out and make contact with the segments of the commutator. This effectively short-circuits all the commutator segments together, and the motor operates in the same manner as a squirrel-cage motor.

The brush-lifting-type motor uses a radial commutator *(Figure 31–45)*. Weights are mounted at the front of the armature. When the motor reaches about 75% of full speed, these weights swing outward due to centrifugal force and cause two push rods to act against a spring barrel and short-circuiting necklace. The weights overcome the force of the spring and cause the entire spring barrel and brush holder assembly to move toward the back of the motor

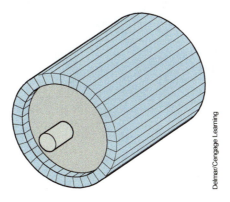

FIGURE 31–43 Axial commutator.

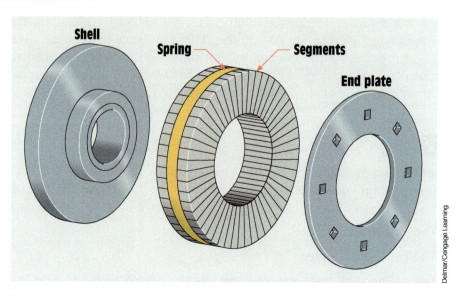

FIGURE 31–44 Short-circuiting ring for brush-riding-type repulsion-start induction-run motor.

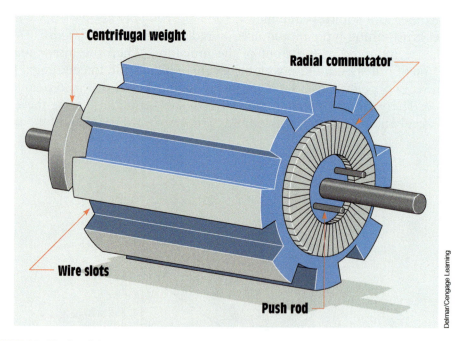

FIGURE 31–45 A radial commutator is used with the brush-lifting-type motor.

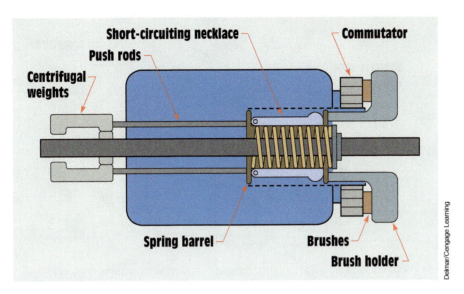

FIGURE 31–46 Brush-lifting-type repulsion-start induction-run motor.

(Figure 31–46). The motor is so designed that the short-circuiting necklace will short-circuit the commutator bars before the brushes lift off the surface of the radial commutator. The motor will now operate as a squirrel-cage induction motor. The brush-lifting motor has several advantages over the brush-riding motor. Because the brushes lift away from the commutator surface during operation, wear on both the commutator and brushes is greatly reduced. Also, the motor does not have to overcome the friction of the brushes riding against the commutator surface during operation. As a result, the brush-lifting motor is quieter in operation.

31–13 Repulsion-Induction Motors

The repulsion-induction motor is basically the same as the repulsion motor except that a set of squirrel-cage windings are added to the armature *(Figure 31–47)*. This type of motor contains no centrifugal mechanism or short-circuiting device. The brushes ride against the commutator at all times. The repulsion-induction motor has very high starting torque because it starts as a repulsion motor. The squirrel-cage winding, however, gives it much better speed characteristics than a standard repulsion motor. This motor has very good speed regulation between no load and full load. Its running characteristics are similar to a DC compound motor. The schematic symbol for a repulsion motor is shown in *Figure 31–48*.

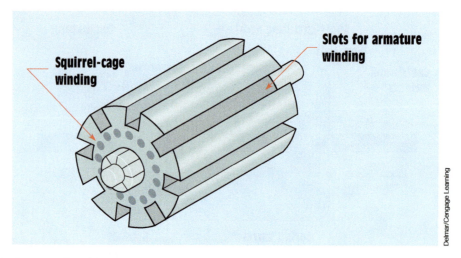

FIGURE 31–47 Repulsion-induction motors contain both armature and squirrel-cage windings.

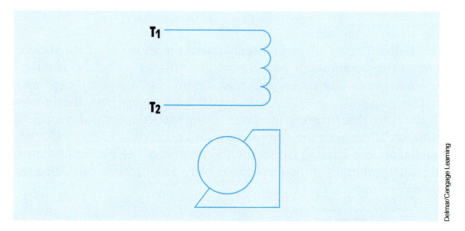

FIGURE 31–48 Schematic symbol for a repulsion motor.

31–14 Single-Phase Synchronous Motors

Single-phase **synchronous motors** are small and develop only fractional horsepower. They operate on the principle of a rotating magnetic field developed by a shaded-pole stator. Although they will operate at synchronous speed, they do not require DC excitation. They are used in applications where constant speed is required such as clock motors, timers, and recording instruments. They also are used as the driving force for small fans because they are small and inexpensive to manufacture. There are two basic types of synchronous motor: the Warren, or General Electric motor, and the Holtz motor. These motors are also referred to as hysteresis motors.

Warren Motors

The **Warren motor** is constructed with a laminated stator core and a single coil. The coil is generally wound for 120-VAC operation. The core contains two poles, which are divided into two sections each. One half of each pole piece contains a shading coil to produce a rotating magnetic field *(Figure 31–49)*. Because the stator is divided into two poles, the synchronous field speed is 3600 rpm when connected to 60 hertz.

The difference between the Warren and Holtz motor is the type of rotor used. The rotor of the Warren motor is constructed by stacking hardened steel laminations onto the rotor shaft. These disks have high hysteresis loss. The laminations form two crossbars for the rotor. When power is connected to the motor, the rotating magnetic field induces a voltage into the rotor and a strong starting torque is developed causing the rotor to accelerate to near-synchronous

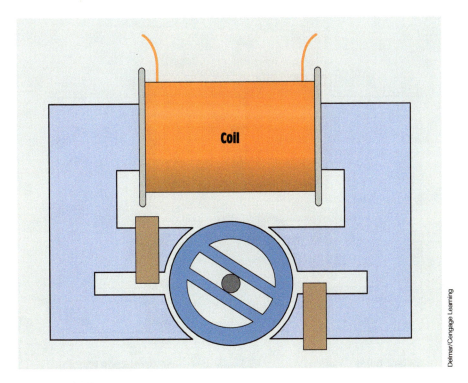

FIGURE 31–49 A Warren motor.

speed. Once the motor has accelerated to near-synchronous speed, the flux of the rotating magnetic field follows the path of minimum reluctance (magnetic resistance) through the two crossbars. This causes the rotor to lock in step with the rotating magnetic field, and the motor operates at 3600 rpm. These motors are often used with small geartrains to reduce the speed to the desired level.

Holtz Motors

The **Holtz motor** uses a different type of rotor *(Figure 31–50)*. This rotor is cut in such a manner that six slots are formed. These slots form six salient (projecting or jutting) poles for the rotor. A squirrel-cage winding is constructed by inserting a metal bar at the bottom of each slot. When power is connected to the motor, the squirrel-cage winding provides the torque necessary to start the rotor turning. When the rotor approaches synchronous speed, the salient poles lock in step with the field poles each half cycle. This produces a rotor speed of 1200 rpm (one-third of synchronous speed) for the motor.

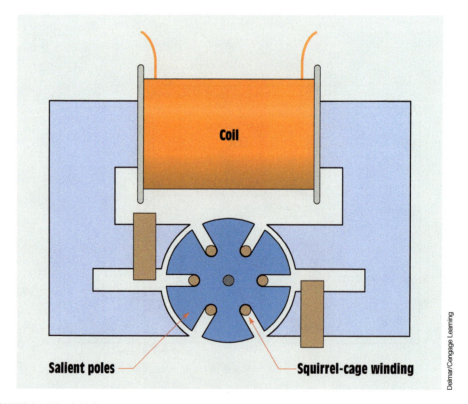

FIGURE 31–50 A Holtz motor.

31–15 Stepping Motors

Stepping motors are devices that convert electric impulses into mechanical movement. Stepping motors differ from other types of DC or AC motors in that their output shaft moves through a specific angular rotation each time the motor receives a pulse. The stepping motor allows a load to be controlled as to speed, distance, or position. These motors are very accurate in their control performance. There is generally less than 5% error per angle of rotation, and this error is not cumulative regardless of the number of rotations. Stepping motors are operated on DC power but can be used as a two-phase synchronous motor when connected to AC power.

Theory of Operation

Stepping motors operate on the theory that like magnetic poles repel and unlike magnetic poles attract. Consider the circuit shown in *Figure 31–51*. In this illustration, the rotor is a permanent magnet and the stator windings consist of two electromagnets. If current flows through the winding of stator pole A in such a direction that it creates a north magnetic pole and through B in such a direction that it creates a south magnetic pole, it is impossible to determine the direction of rotation. In this condition, the rotor could turn in either direction.

Now consider the circuit shown in *Figure 31–52*. In this circuit, the motor contains four stator poles instead of two. The direction of current flow through stator pole A is still in such a direction as to produce a north magnetic field,

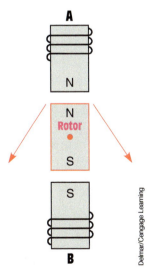

FIGURE 31–51 The rotor could turn in either direction.

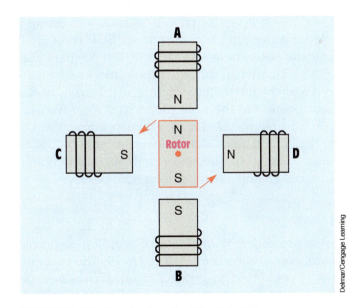

FIGURE 31–52 The direction of rotation is known.

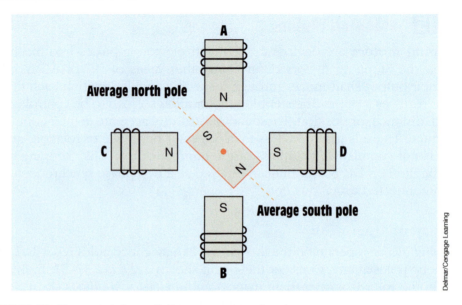

FIGURE 31–53   The magnet aligns with the average magnetic pole.

and the current flow through pole B produces a south magnetic field. The current flow through stator pole C, however, produces a south magnetic field and the current flow through pole D produces a north magnetic field. In this illustration, there is no doubt as to the direction or angle of rotation. In this example, the rotor shaft turns 90° in a counterclockwise direction.

Figure 31–53 shows yet another condition. In this example, the current flow through Poles A and C is in such a direction as to form a north magnetic pole, and the direction of current flow through Poles B and D forms south magnetic poles. In this illustration, the permanent magnetic rotor has rotated to a position between the actual pole pieces.

To allow for better stepping resolution, most stepping motors have eight stator poles, and the pole pieces and rotor have teeth machined into them, as shown in *Figure 31–54*. In actual practice, the number of teeth machined in the stator and rotor determines the angular rotation achieved each time the motor is stepped. The stator-rotor tooth configuration shown in *Figure 31–54* produces an angular rotation of 1.8° per step.

Windings

There are different methods of winding stepping motors. A standard three-lead motor is shown in *Figure 31–55*. The common terminal of the two windings is connected to ground of an above- and below-ground power supply. Terminal 1 is connected to the common of a single-pole double-throw switch (Switch 1), and Terminal 3 is connected to the common of another single-pole

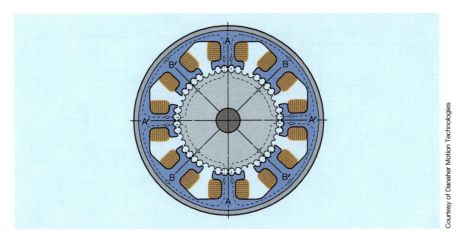

FIGURE 31–54 Construction of a stepping motor.

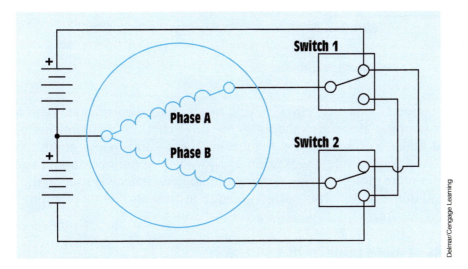

FIGURE 31–55 A standard three-lead motor.

double-throw switch (Switch 2). One of the stationary contacts of each switch is connected to the positive, or above-ground, voltage, and the other stationary contact is connected to the negative, or below-ground, voltage. The polarity of each winding is determined by the position setting of its control switch.

Stepping motors can also be wound bifilar, as shown in *Figure 31–56*. The term *bifilar* means that two windings are wound together. This is similar to a transformer winding with a center-tap lead. Bifilar stepping motors have twice as many windings as the three-lead type, which makes it necessary to use smaller wire in the windings. This results in higher wire resistance in the winding, producing a better inductive-resistive (LR) time constant for the bifilar-wound motor.

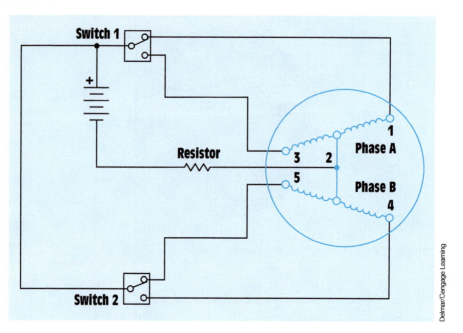

FIGURE 31-56 Bifilar-wound stepping motor.

The increased LR time constant results in better motor performance. The use of a bifilar stepping motor also simplifies the drive circuitry requirements. Notice that the bifilar motor does not require an above- and below-ground power supply. As a general rule, the power supply voltage should be about five times greater than the motor voltage. A current-limiting resistance is used in the common lead of the motor. This current-limiting resistor also helps to improve the LR time constant.

Four-Step Switching (Full-Stepping)

The switching arrangement shown in *Figure 31–56* can be used for a four-step switching sequence (full-stepping). Each time one of the switches changes position, the rotor advances one-fourth of a tooth. After four steps, the rotor has turned the angular rotation of one "full" tooth. If the rotor and stator have 50 teeth, it will require 200 steps for the motor to rotate one full revolution. This corresponds to an angular rotation of 1.8° per step (360°/200 steps = 1.8° per step). The chart shown in *Table 31–1* illustrates the switch positions for each step.

Eight-Step Switching (Half-Stepping)

Figure 31–57 illustrates the connections for an eight-step switching sequence (half-stepping). In this arrangement, the center-tap leads for Phases A and B

Step	Switch 1	Switch 2
1	1	5
2	1	4
3	3	4
4	3	5
1	1	5

TABLE 31–1 Four-Step Switching Sequence

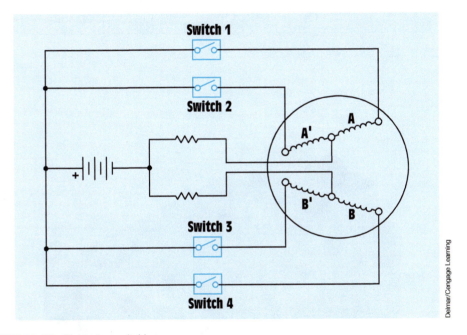

FIGURE 31–57 Eight-step switching.

are connected through their own separate current-limiting resistors back to the negative of the power supply. This circuit contains four separate single-pole switches instead of two switches. The advantage of this arrangement is that each step causes the motor to rotate one-eighth of a tooth instead of one-fourth of a tooth. The motor now requires 400 steps to produce one revolution, which produces an angular rotation of 0.9° per step. This results in better stepping resolution and greater speed capability. The chart in *Table 31–2* illustrates the switch position for each step. A stepping motor is shown in *Figure 31–58*.

974 SECTION XIV AC Machines

Step	Switch 1	Switch 2	Switch 3	Switch 4
1	On	Off	On	Off
2	On	Off	Off	Off
3	On	Off	Off	On
4	Off	Off	Off	On
5	Off	On	Off	On
6	Off	On	Off	Off
7	Off	On	On	Off
8	Off	Off	On	Off
1	On	Off	On	Off

TABLE 31–2 Eight-Step Switching Sequence

FIGURE 31–58 Stepping motor.

AC Operation

Stepping motors can be operated on AC voltage. In this mode of operation, they become two-phase, AC, synchronous, constant-speed motors and are classified as *permanent magnet induction motors*. Refer to the cutaway of a stepping motor shown in *Figure 31–59*. Notice that this motor has no brushes, sliprings, commutator, gears, or belts. Bearings maintain a constant air gap between the permanent magnet rotor and the stator windings. A typical eight-stator-pole stepping motor will have a synchronous speed of 72 rpm when connected to a 60-hertz, two-phase AC powerline.

A resistive-capacitive network can be used to provide the 90° phase shift needed to change single-phase AC into two-phase AC. A simple forward-off-reverse switch can be added to provide directional control. A sample circuit

FIGURE 31–59 Cutaway of a stepping motor.

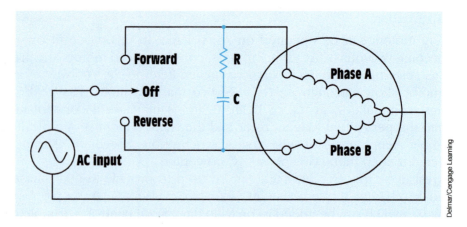

FIGURE 31–60 Phase-shift circuit converts single-phase into two-phase.

of this type is shown in *Figure 31–60*. The correct values of resistance and capacitance are necessary for proper operation. Incorrect values can result in random direction of rotation when the motor is started, change of direction when the load is varied, erratic and unstable operation, and failure to start. The correct values of resistance and capacitance will be different with different stepping motors. The manufacturer's recommendations should be followed for the particular type of stepping motor used.

Stepping Motor Characteristics

When stepping motors are used as two-phase synchronous motors, they can start, stop, or reverse direction of rotation virtually instantly. The motor will start within about 1-1/2 cycles of the applied voltage and will stop within 5 to 25 milliseconds. The motor can maintain a stalled condition without harm to the motor. Because the rotor is a permanent magnet, there is no induced current in the rotor. There is no high inrush of current when the motor is started. The starting and running currents are the same. This simplifies the power requirements of the circuit used to supply the motor. Due to the permanent magnetic structure of the rotor, the motor does provide holding torque when turned off. If more holding torque is needed, DC voltage can be applied to one or both windings when the motor is turned off. An example circuit of this type is shown in *Figure 31–61*. If DC is applied to one winding, the holding torque will be approximately 20% greater than the rated torque of the motor. If DC is applied to both windings, the holding torque will be about 1 to 1/2 times greater than the rated torque.

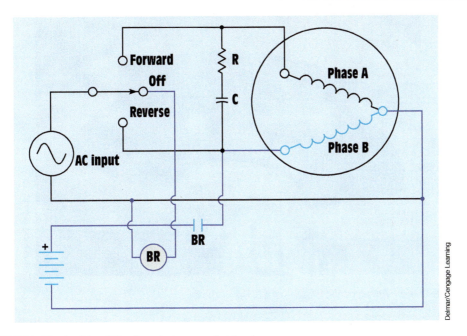

FIGURE 31-61 Applying DC voltage to increase holding torque.

31-16 Universal Motors

The **universal motor** is often referred to as an AC series motor. This motor is very similar to a DC series motor in its construction in that it contains a wound armature and brushes *(Figure 31-62)*. The universal motor, however, has the addition of a **compensating winding.** If a DC series motor is connected to AC, the motor operates poorly for several reasons. The armature windings have a large amount of inductive reactance when connected to AC. Another reason for poor operation is that the field poles of most DC machines contain solid metal pole pieces. If the field is connected to AC, a large amount of power is lost to eddy current induction in the pole pieces. Universal motors contain a laminated core to help prevent this problem. The compensating winding is wound around the stator and functions to counteract the inductive reactance in the armature winding.

The universal motor is so named because it can be operated on AC or DC voltage. When the motor is operated on DC, the compensating winding is connected in series with the series field winding *(Figure 31-63)*.

Connecting the Compensating Winding for AC

When the universal motor is operated with AC power, the compensating winding can be connected in two ways. If it is connected in series

FIGURE 31–62 Armature and brushes of a universal motor.

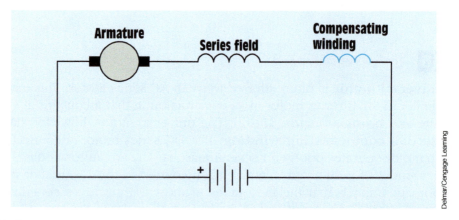

FIGURE 31–63 The compensating winding is connected in series with the series field winding.

with the armature, as shown in *Figure 31–64*, it is known as **conductive compensation.**

The compensating winding can also be connected by shorting its leads together, as shown in *Figure 31–65*. When connected in this manner, the winding acts like a shorted secondary winding of a transformer. Induced current permits the winding to operate when connected in this manner. This connection is known as **inductive compensation.** Inductive compensation cannot be used when the motor is connected to DC.

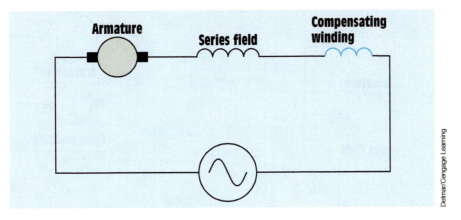

FIGURE 31–64 Conductive compensation.

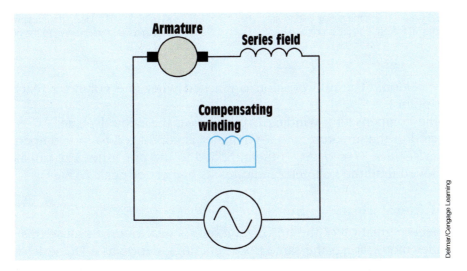

FIGURE 31–65 Inductive compensation.

The Neutral Plane

Because the universal motor contains a wound armature, commutator, and brushes, the brushes should be set at the **neutral plane** position. This can be done in the universal motor in a manner similar to that of setting the neutral plane of a DC machine. When setting the brushes to the neutral plane position in a universal motor, either the series or compensating winding can be used. To set the brushes to the neutral plane position using the series winding *(Figure 31–66)*, AC is connected to the armature leads. A voltmeter is connected to the series winding. Voltage is then applied to the armature. The brush position is then moved until the voltmeter connected to the series field reaches

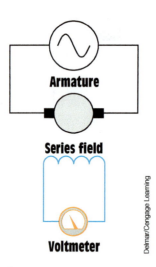

FIGURE 31–66 Using the series field to set the brushes at the neutral plane position.

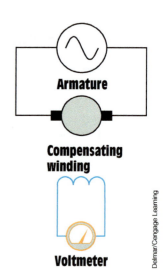

FIGURE 31–67 Using the compensating winding to set the brushes to the neutral plane position.

a null position. (The null position is reached when the voltmeter reaches its lowest point.)

If the compensating winding is used to set the neutral plane, AC is again connected to the armature and a voltmeter is connected to the compensating winding *(Figure 31–67)*. AC is then applied to the armature. The brushes are then moved until the voltmeter indicates its highest or peak voltage.

Speed Regulation

The speed regulation of the universal motor is very poor. Because this motor is a series motor, it has the same poor speed regulation as a DC series motor. If the universal motor is connected to a light load or no load, its speed is almost unlimited. It is not unusual for this motor to be operated at several thousand revolutions per minute. Universal motors are used in a number of portable appliances where high horsepower and light weight are needed, such as drill motors, skill saws, and vacuum cleaners. The universal motor is able to produce a high horsepower for its size and weight because of its high operating speed.

Changing the Direction of Rotation

The direction of rotation of the universal motor can be changed in the same manner as changing the direction of rotation of a DC series motor. To change the direction of rotation, change the armature leads with respect to the field leads.

Summary

- Not all single-phase motors operate on the principle of a rotating magnetic field.
- Split-phase motors start as two-phase motors by producing an out-of-phase condition for the current in the run winding and the current in the start winding.
- The resistance of the wire in the start winding of a resistance-start induction-run motor is used to produce a phase angle difference between the current in the start winding and the current in the run winding.
- The capacitor-start induction-run motor uses an AC electrolytic capacitor to increase the phase angle difference between starting and running current. This causes an increase in starting torque.
- Maximum starting torque for a split-phase motor is developed when the start-winding current and run-winding current are 90° out of phase with each other.
- Most resistance-start induction-run motors and capacitor-start induction-run motors use a centrifugal switch to disconnect the start windings when the motor reaches approximately 75% of full-load speed.
- The capacitor-start capacitor-run motor operates like a two-phase motor because both the start and run windings remain energized during motor operation.
- Most capacitor-start capacitor-run motors use an AC oil-filled capacitor connected in series with the start winding.
- The capacitor of the capacitor-start capacitor-run motor does help to correct the power factor.
- Shaded-pole induction motors operate on the principle of a rotating magnetic field.
- The rotating magnetic field of a shaded-pole induction motor is produced by placing shading loops or coils on one side of the pole piece.
- The synchronous-field speed of a single-phase motor is determined by the number of stator poles and the frequency of the applied voltage.
- Consequent-pole motors are used when a change of motor speed is desired and high torque must be maintained.
- Multispeed fan motors are constructed by connecting windings in series with the main run winding.

- Multispeed fan motors have high-impedance stator windings to prevent them from overheating when their speed is reduced.
- There are three basic repulsion-type motors: the repulsion motor, the repulsion-start induction-run motor, and the repulsion-induction motor.
- Repulsion motors have the highest starting torque of any single-phase motor.
- The direction of rotation of repulsion motors is changed by setting the brushes 15° on either side of the pole pieces.
- The direction of rotation for split-phase motors is changed by reversing the start winding in relation to the run winding.
- Shaded-pole motors are generally considered to be nonreversible.
- There are two types of repulsion-start induction-run motors: the brush-riding type and the brush-lifting type.
- The brush-riding type of motor uses an axial commutator and a short-circuiting device, which short-circuits the commutator segments when the motor reaches approximately 75% of full-load speed.
- The brush-lifting type of repulsion-start induction-run motor uses a radial commutator. A centrifugal device causes the brushes to move away from the commutator and a short-circuiting necklace to short-circuit the commutator when the motor reaches about 75% of full-load speed.
- The repulsion-induction motor contains both a wound armature and squirrel-cage windings.
- There are two types of single-phase synchronous motor: the Warren and the Holtz.
- Single-phase synchronous motors are sometimes called hysteresis motors.
- The Warren motor operates at a speed of 3600 rpm.
- The Holtz motor operates at a speed of 1200 rpm.
- Stepping motors generally operate on DC and are used to produce angular movements in steps.
- Stepping motors are generally used for position control.
- Stepping motors can be used as synchronous motors when connected to two-phase AC.
- Stepping motors operate at a speed of 72 rpm when connected to 60-hertz power.

- Stepping motors can produce a holding torque when DC is connected to their windings.
- Universal motors operate on DC or AC.
- Universal motors contain a wound armature and brushes.
- Universal motors are also called AC series motors.
- Universal motors have a compensating winding that helps overcome inductive reactance.
- The direction of rotation for a universal motor can be changed by reversing the armature leads with respect to the field leads.

Review Questions

1. What are the three basic types of split-phase motors?
2. The voltages of a two-phase system are how many degrees out of phase with each other?
3. How are the start and run windings of a split-phase motor connected in relation to each other?
4. In order to produce maximum starting torque in a split-phase motor, how many degrees out of phase should the start- and run-winding currents be with each other?
5. What is the advantage of the capacitor-start induction-run motor over the resistance-start induction-run motor?
6. On the average, how many degrees out of phase with each other are the start- and run-winding currents in a resistance-start induction-run motor?
7. What device is used to disconnect the start windings for the circuit in most nonhermetically sealed capacitor-start induction-run motors?
8. Why does a split-phase motor continue to operate after the start windings have been disconnected from the circuit?
9. How can the direction of rotation of a split-phase motor be reversed?
10. If a dual-voltage split-phase motor is to be operated on high voltage, how are the run windings connected in relation to each other?
11. When determining the direction of rotation for a split-phase motor, should you face the motor from the front or from the rear?

12. What type of split-phase motor does not generally contain a centrifugal switch?

13. What type of single-phase motor develops the highest starting torque?

14. What is the principle of operation of a repulsion motor?

15. What type of commutator is used with a brush-lifting-type repulsion-start induction-run motor?

16. When a repulsion-start induction-run motor reaches about 75% of rated full-load speed, it stops operating as a repulsion motor and starts operating as a squirrel-cage motor. What must be done to cause the motor to begin operating as a squirrel-cage motor?

17. What is the principle of operation of a capacitor-start capacitor-run motor?

18. What causes the magnetic field to rotate in a shaded-pole induction motor?

19. How can the direction of rotation of a shaded-pole induction motor be changed?

20. How is the speed of a consequent-pole motor changed?

21. Why can a multispeed fan motor be operated at lower speed than most induction motors without harm to the motor windings?

22. What is the speed of operation of the Warren motor?

23. What is the speed of operation of the Holtz motor?

24. Explain the difference in operation between a stepping motor and a common DC motor.

25. What is the principle of operation of a stepping motor?

26. What does the term *bifilar* mean?

27. Why do stepping motors have teeth machined in the stator poles and rotor?

28. When a stepping motor is connected to AC power, how many phases must be applied to the motor?

29. What is the synchronous speed of an eight-pole stepping motor when connected to a two-phase, 60-Hz AC line?

30. How can the holding torque of a stepping motor be increased?

31. Why is the AC series motor often referred to as a universal motor?

32. What is the function of the compensating winding?

33. How is the direction of rotation of the universal motor reversed?
34. When the motor is connected to DC voltage, how must the compensating winding be connected?
35. Explain how to set the neutral plane position of the brushes using the series field.
36. Explain how to set the neutral plane position using the compensating winding.

Practical Applications

You are an electrical contractor, and you have been called to a home to install a well pump. The homeowner has purchased the pump but does not know how to connect it. You open the connection terminal cover and discover that the motor contains eight terminal leads marked T_1 through T_8. The motor is to be connected to 240 V. At present, the T leads are connected as follows: T_1, T_3, T_5, and T_7 are connected together; and T_2, T_4, T_6, and T_8 are connected together. L_1 is connected to the group of terminals with T_1, and L_2 is connected to the group of terminals with T_2. Is it necessary to change the leads for operation on 240 V? If so, how should they be connected? ■

APPENDIX F

Answers to Practice Problems

Unit 26 Three-Phase Circuits

1.
   $E_{P(A)}$ 138.568 V $E_{P(L)}$ 240 V
   $I_{P(A)}$ 34.64 A $I_{P(L)}$ 20 A
   $E_{L(A)}$ 240 V $E_{L(L)}$ 240 V
   $I_{L(A)}$ 34.64 A $I_{L(L)}$ 34.64 A
   P 14,399.155 W Z_{PHASE} 12 Ω

2.
   $E_{P(A)}$ 4160 V $E_{P(L)}$ 2401.849 V
   $E_{P(a)}$ 23.112 A $E_{P(L)}$ 40.031 A
   $E_{L(A)}$ 4160 V $E_{L(L)}$ 4160 V
   $I_{L(A)}$ 40.031 A $I_{L(L)}$ 40.031 A
   P 288,428.159 W Z_{PHASE} 60 Ω

3.
   $E_{P(A)}$ 323.326 V $E_{P(L1)}$ 323.326 V $E_{P(L2)}$ 560 V
   $I_{P(A)}$ 185.91 A $I_{P(L1)}$ 64.665 A $I_{P(L2)}$ 70 A
   $E_{L(A)}$ 560 V $E_{L(L1)}$ 560 V $E_{L(L2)}$ 560 V
   $I_{L(A)}$ 185.905 A $I_{L(L1)}$ 64.665 A $I_{L(L2)}$ 121.24 A
   P 180,317.83 W Z_{PHASE} 5 Ω Z_{PHASE} 8 Ω

4.
   $E_{P(A)}$ 277.136 V $E_{P(L1)}$ 277.136 V $E_{P(L2)}$ 480 V $E_{P(L3)}$ 277.136 V
   $I_{P(A)}$ 34.485 A $I_{P(L1)}$ 23.095 A $I_{P(L2)}$ 30 A $I_{P(L3)}$ 27.714 A
   $E_{L(A)}$ 480 V $E_{L(L1)}$ 480 V $E_{L(L2)}$ 480 V $E_{L(L3)}$ 480 V
   $I_{L(A)}$ 34.049 A $I_{L(L1)}$ 23.095 A $I_{L(L2)}$ 51.96 A $I_{L(L3)}$ 27.714 A
   VA 27,838.09 Z_{PHASE} 12 Ω $Z_{L(PHASE)}$ 16 Ω $X_{C(PHASE)}$ 10 Ω
                          P 19,200.259 W $VARs_L$ 43,197.466 $VARs_C$ 23,040.311

Unit 27 Single-Phase Transformers

1.
   - E_P 120 V
   - I_P 1.6 A
   - N_P 300 turns
   - Ratio 5:1
   - E_S 24 V
   - I_S 8 A
   - N_S 60 turns
   - $Z = 3\ \Omega$

2.
   - E_P 240 V
   - I_P 0.853 A
   - N_P 210 turns
   - Ratio 1:1.333
   - E_S 320 V
   - I_S 0.643 A
   - N_S 280 turns
   - $Z = 500\ \Omega$

3.
   - E_P 64 V
   - I_P 33.333 A
   - N_P 32 turns
   - Ratio 1:2.5
   - E_S 160 V
   - I_S 13.333 A
   - N_S 80 turns
   - $Z = 12\ \Omega$

4.
   - E_P 48 V
   - I_P 3.333 A
   - N_P 220 turns
   - Ratio 1:5
   - E_S 240 V
   - I_S 0.667 A
   - N_S 1100 turns
   - $Z = 360\ \Omega$

5.
   - E_P 35.848 V
   - I_P 16.5 A
   - N_P 87 turns
   - Ratio 1:5.077
   - E_S 182 V
   - I_S 3.25 A
   - N_S 450 turns
   - $Z = 56\ \Omega$

6.
   - E_P 480 V
   - I_P 1.458 A
   - N_P 275 turns
   - Ratio 1:1.909
   - E_S 916.346 V
   - I_S 0.764 A
   - N_S 525 turns
   - $Z = 1.2\ k\Omega$

7.
   - E_P 208 V
   - I_P 11.93 A
   - N_P 800 turns
   - E_{S1} 320 V
   - I_{S1} 0.0267 A
   - N_{S1} 1231 turns
   - Ratio 1 1:1.54
   - R_1 12 kΩ
   - E_{S2} 120 V
   - I_{S2} 20 A
   - N_{S2} 462 turns
   - Ratio 2 1.73:1
   - R_2 6 Ω
   - E_{S3} 24 V
   - I_{S3} 3 A
   - N_{S3} 92 turns
   - Ratio 3 1:8.67
   - R_3 8 Ω

8.

E_P 277 V $\qquad$ E_{S1} 480 V $\qquad$ E_{S2} 208 V $\qquad$ E_{S3} 120 V
I_P 8.93 A $\qquad$ I_{S1} 2.4 A $\qquad$ I_{S2} 3.47 A $\qquad$ I_{S3} 5 A
N_P 350 turns $\qquad$ N_{S1} 606 turns $\qquad$ N_{S2} 263 turns $\qquad$ N_{S3} 152 turns
$\qquad\qquad\qquad$ Ratio 1 1:1.73 $\qquad$ Ratio 2 1.33:1 $\qquad$ Ratio 3 2.31:1
$\qquad\qquad\qquad$ R_1 200 Ω $\qquad$ R_2 60 Ω $\qquad$ R_3 24 Ω

Unit 28 Three-Phase Transformers

1.

E_P 2401.8 V $\qquad$ E_P 440 V $\qquad$ E_P 254.04 V
I_P 7.67 A $\qquad$ I_P 41.9 A $\qquad$ I_P 72.58 A
E_L 4160 V $\qquad$ E_L 440 V $\qquad$ E_L 440 V
I_L 7.67 A $\qquad$ I_L 72.58 A $\qquad$ I_L 72.58 A
Ratio 5.46:1 $\qquad$ Z 3.5 Ω

2.

E_P 4157.04 V $\qquad$ E_P 240 V $\qquad$ E_P 138.57 V
I_P 1.15 A $\qquad$ I_P 20 A $\qquad$ I_P 34.64 A
E_L 7200 V $\qquad$ E_L 240 V $\qquad$ E_L 240 V
I_L 1.15 A $\qquad$ I_L 34.64 A $\qquad$ I_L 34.64 A
Ratio 17.32:1 $\qquad$ Z 4 Ω

3.

E_P 13,800 V $\qquad$ E_P 277 V $\qquad$ E_P 480 V
I_P 6.68 A $\qquad$ I_P 332.54 A $\qquad$ I_P 192 A
E_L 13,800 V $\qquad$ E_L 480 V $\qquad$ E_L 480 V
I_L 11.57 A $\qquad$ I_L 332.54 A $\qquad$ I_L 332.54 A
Ratio 49.76:1 $\qquad$ Z 2.5 Ω

4.

E_P 23,000 V $\qquad$ E_P 120 V $\qquad$ E_P 208 V
I_P 0.626 A $\qquad$ I_P 120.08 A $\qquad$ I_P 69.33 A
E_L 23,000 V $\qquad$ E_L 208 V $\qquad$ E_L 208 V
I_L 1.08 A $\qquad$ I_L 120.08 A $\qquad$ I_L 120.08 A
Ratio 191.66:1 $\qquad$ Z 3 Ω

INDEX

Alternator calculations, 749–750
Alternator cooling, 865
Amortisseur winding, 919
Autotransformers, 788–791

Bifilar, 971
Brushless exciter, 862–864, 917

Capacitor-start capacitor-run motors, 950–953
Capacitor-start induction-run motors, 944–945
Centrifugal switch, 937
Closed delta
    with center tap, 834
    without center tap, 835–836
Closing a delta, 822–823
    polarity testing, 822–823
Code letters, 904–905
Compensating winding, 977
Conductive compensation, 977–978
Consequent-pole, 910–913
Consequent-pole motor, 957–958
Constant-current transformer, 804–806
Control transformer, 778–780
Current-limiting resistance, 972
Current regulator, 804
Current relay, 940–941

Delta connection, 736
Delta–wye, 819
    connection with neutral, 836–837
Dielectric oil, 817
Differential selsyn, 925–926
Direction of rotation, 887–889
Distribution transformer, 775–778
Dual-voltage motors, 889
Dual-voltage split-phase motors, 946–949
Dual-voltage three-phase motors, 889–895

high-voltage connections, 890
low-voltage connections, 890–893
voltage and current relationships for, 894–895
Dynamic-type wattmeter, 725

Electronic-type wattmeter, 725
Excitation current, 764

Field-discharge resistor, 871, 919
Flux leakage, 784
Frequency, 866–867
    determining factors of, 867

Harmonic derating factor, 850–851
Harmonic problems, 846–850
Harmonics, 842–851
    dealing with, 849–850
    derating factor determination, 850–851
    effects of, 843–846
    problems determination on single-phase systems, 846–848
    problems determination on three-phase systems, 848–849
High leg, 832
High-voltage connections, 890
Holtz motor, 968
Hot-wire relay, 939–940
H-type core, 782
Hydrogen, 865

Inductive compensation, 978
Inrush current, 784
Interconnected-wye, 840
Isolation transformers, 762–788
    control transformer, 778–780
    core types, 781–784
    distribution transformer, 775–778
    excitation current, 764
    inrush current, 784

magnetic domains, 784–788
multiple-tapped windings, 770–772
mutual induction, 764–765
operating principles of, 764
transformer calculations, 765–767
value calculation using turns ratio, 768–770
value calculation with multiple secondaries, 772–775

Laminated cores, 781
Leakage current, 942
Line current, 733, 736
Line voltage, 732–733, 736
Load 1 calculations, 749
Load conditions, 832–833
Load sharing, 871
Low-voltage connections, 890–893

Magnetic domains, 784–788
Motor primary, 876
Motor secondary, 876
Multiple-tapped windings, 770–772
Multispeed motors, 957–959
    consequent-pole type, 957–958
    fan motors, 958–959
    shaded-pole induction motor, 957–958
Mutual induction, 764–765

Neutral conductor, 775
Neutral plane, 979–980
Nikola Tesla, 730, 876

One-line diagram, 821–822
Open-delta, 829–830
    connection, 829–830, 831
Orange wire, 832
Output voltage, 867–868
    controlling of, 868
    determining factors of, 867–868

Parallel alternators, 868
Paralleling alternators, 868–870
   conditions for, 868–869
   phase rotation determination, 869–870
   synchroscope, 870
Percent slip, 902
Phase current, 733, 736
Phase rotation, 869–870
Phase rotation meter, 887
Phase voltage, 732–733, 736
Power factor correction, 750–754
Power factor correction, 921
Primary winding, 762, 836
Pullout torque, 920

Receiver, 923–924
Repulsion-induction motors, 965–966
Repulsion motors, 959–963
   brush position, 960–963
   operation, 960
Repulsion-start induction-run motors, 963–965
   brush-lifting-type, 963–965
   brush-riding type, 963
Repulsion-type motors, 959–966
   repulsion-induction motors, 965–966
   repulsion motors, 959–963
   repulsion-start induction-run motors, 963–965
Resistance-start induction-run motor, 936–944
   relationship of stator and rotor fields, 942–943
   rotation direction of, 944
   Starting relays, 937–942
   start winding disconnection, 937
Revolving-armature-type alternator, 858–859
Revolving-field-type alternator, 860–862
Rotating magnetic field, 877–889, 933–935, 953
   synchronous speed, 881–886
Rotor, 862
   frequency, 902–903
Run winding, 933, 936

Scott connection, 840
Secondary winding, 762
Self-induction, 764

Selsyn motors, 923–926
   differential selsyn, 925–926
   operation of, 924–925
Shaded-pole induction motors, 953–958
   operating characteristics, 957
   shading coil, 954–955
   speed, 956
Shading coil, 954–955
Shell-type transformer, 782
Single-phase loads, 830–834
   load conditions, 832–833
   neutral current calculation, 834
   open-delta connection, 831
   voltage values, 832
Single-phase motors, 932
   multispeed motors, 957
Single-phase synchronous motors, 966–968
   Holtz motor, 968
   Warren motor, 967–968
Single-phase transformers, 760–761
Single-phasing, 907–908
Sliprings, 858
Solid-state starting relay, 941–942
Split-phase motors, 932–935
   classifications of, 932–933
   direction of rotation for, 949–950
   stator windings, 933–935
   two-phase system, 933
Squirrel-cage induction motors, 895–914
   code letters, 904–905
   consequent-pole, 910–913
   double-squirrel-cage rotor, 905–906
   motor calculations, 913–914
   nameplate, 909–910
   operating principles of, 896–898
   percent slip, 902
   power factor, 906
   reduced voltage starting, 903–904
   rotor frequency, 902–903
   single-phasing, 907–908
   starting characteristics of, 899–901
   torque, 898–899
   voltage variation effects on, 908
Squirrel-cage rotor, 895, 953
Starting relays, 937–942
   current relay, type of, 940–941
   hot-wire relay, type of, 939–940
   solid-state starting relay, type of, 941–942

Start winding, 933, 936
Stator, 860–861
   winding, 861, 876, 883, 914, 933–935
Step-down transformer, 766
Stepping motors, 969–977
   AC operation, 975–976
   characteristics of, 976
   eight-step switching (half-stepping), 972–974
   four-step switching (full-stepping), 972
   operating theory of, 969–970
   windings, 970–972
Step-up transformer, 767
Synchronous condenser, 922
   advantages of, 922
Synchronous motors, 917–923, 966–967
   applications of, 922
   constant-speed operation, 919–920
   DC and AC fields interaction, 921
   field-discharge resistor, 919
   power factor correction, 921
   power supply, 920
   rotor construction, 917
   starting, 917–919
Synchronous speed, 881–886
Synchroscope, 870

Tagging, 832
Tape-wound core, 782
T-connected transformers, 837–840
T connection, 837–840
Thermistor, 942
Three-phase alternators, 858–861
   field-discharge protection, 871–872
   load sharing, 871
   output frequency of, 866–867
   output voltage of, 867–868
   revolving-armature-type, 858–860
   revolving-field-type, 860–862
Three-phase bank, 819–820
Three-phase circuits, 730–732
   calculations, 739–747
Three-phase motors, 876
   direction of rotation for, 887–889
   operating principle of, 877

squirrel-cage induction, 895–914
synchronous motors, 917–923
types of, 876
wound-rotor induction, 914–917
Three-phase power, 737–738
Three-phase transformers, 817–821
connections of, 819, 823–829
three-phase bank connection, 819–822
Three-phase VARs, 738
Three-phase watts, 738
Toroid core, 782
Torque, 898–899
Transformer, 760
current determination, 800
formulas, 761
nameplate, 799–800
testing of, 798–799
voltage and current relationships in, 796–800

Transformer impedance, 801–810
parallel transformer connections, 808
precautions, 809–811
series connection of transformer secondaries, 807–808
Transformer polarity
additive and subtractive, 793–794
arrows to place dots, 795
markings schematic, 792
Transmitter, 923–924
Turns ratio, 760, 765, 768–770, 773–775
Two-phase system, 933

Universal motors
changing rotation direction, 980
compensating winding for AC, 977–978
neutral plane, 979–980
speed regulation, 980

Voltage values, 832
Voltage variation effects, 908–909
heat rise, 909
voltage unbalances, 908–909
Volts-per-turn ratio, 765

Warren motor, 967–968
Wattmeter, 724–725
Winding, stator, 861, 876, 883, 914
Wound-rotor induction motors, 914–917
operating principles of, 915–916
speed control, 916–917
starting characteristics of, 916
Wound-rotor motor, 914
Wye connection, 732–735
voltage relationships in, 733–735
Wye–delta, 819

Zig-zag connection, 840–841